MUSHROOM
Production and Products

MUSHROOM
Production and Products

Kuntala Bhattacharjee
Anagha Sarma
Rakesh Kumar

2026
Agri Horti Press
New Delhi-110 084

ISBN: 978-93-49690-51-6 (Hardbound)

Published By

Agri Horti Press

New Delhi

E-mail: agrihortipress@gmail.com

Preface

Mushroom: Production and Products provides a comprehensive guide to the cultivation, processing, and utilization of mushrooms, focusing on the diverse and expanding role they play in modern agriculture and food industries. As a highly nutritious and versatile food source, mushrooms have garnered significant attention for their health benefits, culinary appeal, and potential in sustainable farming systems. This book aims to serve as a detailed resource for researchers, students, industry professionals, and enthusiasts interested in mushroom production and its many applications.

The initial chapters introduce the biological and ecological aspects of mushrooms, covering the various species that are cultivated worldwide, from the widely known Agaricus bisporus to specialty varieties such as oyster mushrooms, shiitake, and enoki. The book provides an in-depth exploration of the biology and lifecycle of mushrooms, emphasizing their unique role in ecosystems as decomposers and their complex reproductive strategies. This foundational knowledge is essential for understanding the environmental and physiological factors that influence mushroom growth.

The subsequent chapters delve into the techniques and technologies used in mushroom cultivation, offering detailed guidance on the optimal conditions for growing different species. Topics such as substrate preparation, inoculation methods, environmental control (temperature, humidity, and light), and harvesting techniques are thoroughly covered. Practical advice is provided for small-scale growers as well as larger commercial mushroom producers, with attention to both traditional and innovative methods in the field.

Mushroom production, while relatively simple in concept, requires a precise balance of various factors. This book explores the challenges faced by mushroom growers, such as pest and disease management, contamination prevention, and optimizing yields. It also examines the economic aspects of mushroom farming, including cost-effective production techniques, marketing strategies, and potential profitability in both local and international markets.

In addition to the production side, the book explores the wide array of mushroom products that have become increasingly popular in the food industry. From fresh mushrooms to dried products, sauces, soups, powders, and nutraceuticals, mushrooms offer a wealth of possibilities for value-added processing. The chapters focus on methods for preserving mushrooms, such as drying, freezing, and canning, as well as the development of new and innovative products like mushroom-based meat substitutes, functional foods, and dietary supplements. The health benefits of mushrooms, including their potential role in supporting immune function, reducing cholesterol, and offering antioxidant properties, are also discussed.

The book also highlights the growing importance of sustainable mushroom production practices. Given the increasing demand for plant-based foods and the push for environmentally friendly farming techniques, mushrooms are seen as an ideal crop for promoting sustainability. Their cultivation requires relatively low input resources, and the waste from mushroom production can often be used as compost or in other agricultural processes. The potential for utilizing agricultural by-products as substrates for mushroom growth is also explored, aligning with the broader push for circular economy models in agriculture.

Furthermore, Mushroom: Production and Products addresses the challenges and opportunities in the global mushroom market. It examines the role of mushrooms in different cultures, their culinary significance, and the increasing demand for exotic varieties. Emerging trends in the mushroom industry, including innovations in biotechnology, automation in production, and the integration of mushrooms in plant-based diets, are also discussed.

This book is intended as a valuable resource for individuals involved in mushroom farming, food processing, research, and product development. It combines practical advice with scientific insights, offering a holistic view of mushroom production and its economic and nutritional importance. The aim is to inspire readers to explore the full potential of mushrooms, whether in small-scale enterprises, commercial agriculture, or innovative product development.

We would like to express our sincere appreciation to the mushroom growers, scientists, food technologists, and industry experts whose research and contributions have helped shape this field. It is our hope that this book will contribute to advancing the understanding of mushroom production and products, fostering sustainable practices and encouraging innovation in this ever-growing industry.

Kuntala Bhattacharjee

Anagha Sarma

Rakesh Kumar

Contents

Chapter 1

Introduction

There is decline in income of farmers in traditional agriculture. To ensure that the farmer gets assured income year after year, there is a need for diversification in agriculture, that is, needs to cultivate different type of crops. One of the options is to grow mushrooms which can be grown on straws and other agricultural wastes.

Unlike plants, mushroom cultivation is an indoor activity. It is possible to grow mushrooms in a particular season under natural conditions. It is possible to grow mushrooms throughout the year under controlled conditions. An additional advantage is that very less land is required for growing mushrooms.

You must have seen that after rains many mushrooms appear from nowhere, especially in grassland, near manure heap, dung or rotting straws/ wood. We must be aware that all mushroom occurring in the nature are not edible. At present there is no simple method to differentiate an edible mushroom from non-edible type. Some mushrooms have medicinal value. A few of these are poisonous and thus it is important to cultivate edible mushrooms.

History of Mushroom Cultivation and Present Status

Mushrooms like wood ear, winter mushroom and shiitake were cultivated in China on wood logs more than a thousand year ago while button mushroom was cultivated in France about four centuries ago. But the scientific cultivation of mushrooms started in the beginning of 20th century and in the initial years button mushroom was the most commonly cultivated mushroom. Subsequently, numbers of other mushrooms were brought under cultivation. In 1960, when mushroom production was low, white button mushroom cultivation was about 80 percent of world mushroom production, shiitake contributed 15 % and others were only 5%.

Over decades there has been increase in production of button, but the production of other mushrooms has increased much more as a result of which at present the relative contribution of button in world production is only 15% and it is no more world's number one mushroom. At present shiitake is number one cultivated mushroom (Fig. 1). Six mushrooms, namely shiitake (*Lentinula*), oyster (*Pleurotus*), wood ear (*Auricularia*), button (*Agaricus*), winter mushroom (*Flammulina*) and paddy straw mushroom (*Volvariella*) account for 90% of the total world mushroom production. Today China cultivates around 60 different types of mushrooms and produces around 85% of the total of 40 million tonne mushrooms of the world.

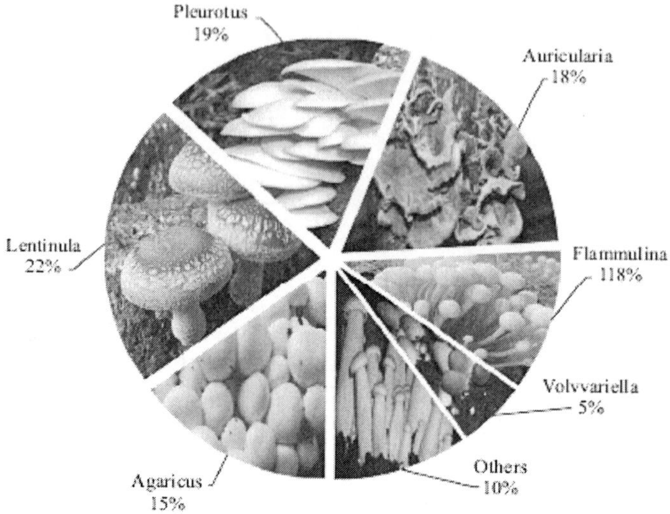

Fig. 1 : Relative contribution of different mushrooms to world mushroom production of 44 million tonne in 2019

In India, we cultivate mainly four types of mushrooms viz., button, oyster, paddy straw and milky mushroom (Fig. 2). The total estimated mushroom production in India in 2019 was around 1.3 lakh tonnes, of which button accounted for 3/4th of the production (Fig 3). Some of the mushrooms like morels are still collected from forests and we have not succeeded in cultivating these in our country.

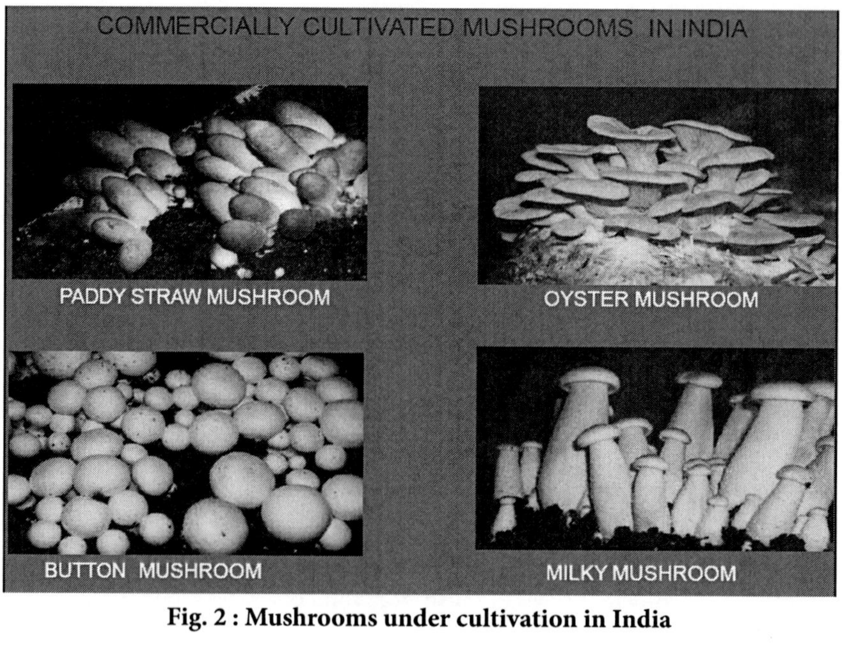

Fig. 2 : Mushrooms under cultivation in India

Button mushroom is an introduced crop in India. You must have listened about other crops like apple, potato that were introduced from other countries to India.

Cultivation of button mushroom in India started in mid 60s in hilly region of Himachal and J&K as the low temperature required was naturally available in these regions. Later on, its cultivation started under controlled conditions in other regions. At present the mushroom production systems in our country are mixed type i.e. both seasonal farming as well as high-tech industry. Mushroom production in India was only 5000 tonnes in 1999 that increased to over 1,00,000 tonnes in 2010, and to 1.30,000 tonne in 2016. At present (2021) the total mushroom production is estimated to be 1,90,000 tonne.

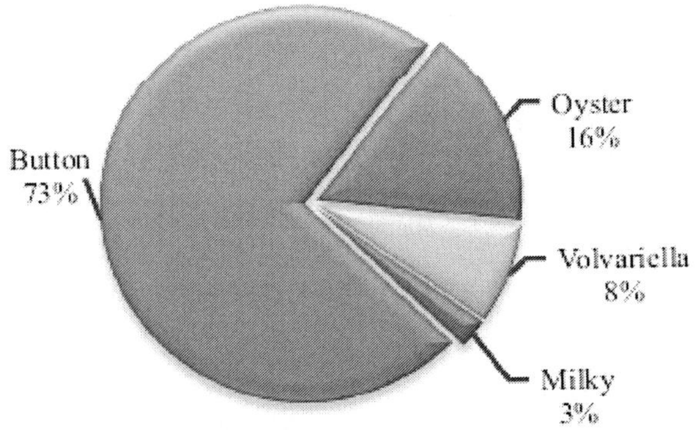

Fig. 3 : Contribution of different species in total mushroom production in India

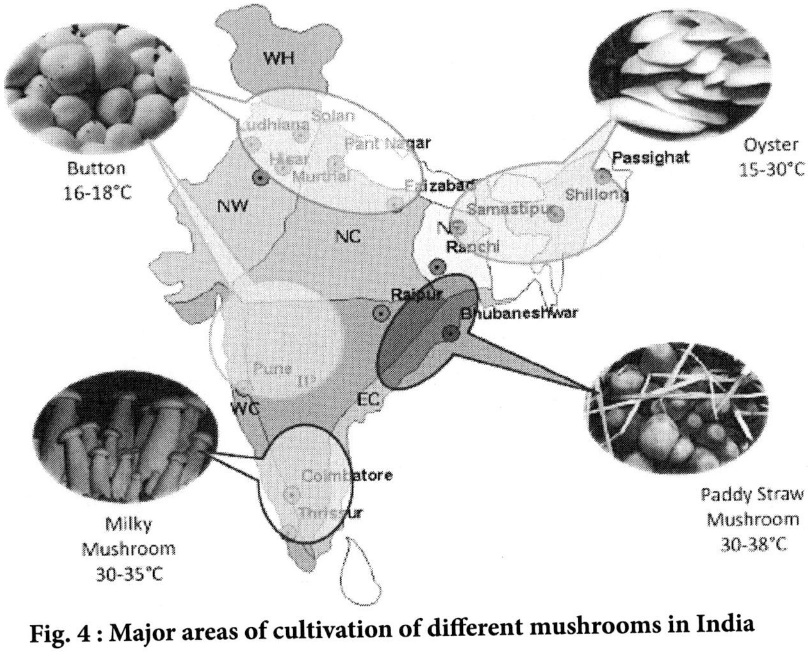

Fig. 4 : Major areas of cultivation of different mushrooms in India

Button mushroom is cultivated throughout the year by commercial units and also during winter months by seasonal growers. The cultivation is more in Maharashtra, Gujarat, Goa, Haryana, Punjab, Uttrakhand, and Himachal Pradesh. Cultivation of paddy straw mushroom is localized to Odisha while milky mushroom is more popular in southern part of India. Oyster is more popular in East, even though it is cultivated in many other parts of the country. There are many species of oyster and globally maximum numbers of species of this mushroom are under cultivation. The regions of mushrooms and the temperature range at which these are cultivated is shown in Fig. 4.

Why to Grow Mushrooms?

In nineteenth century, we produced the food only for our survival. In the last century, we produced adequate food in many parts of the world and convenience was an important area. For example, the ready-to-cook, ready-to-eat foods were commonly available. The 21st century is going to be a century of functional foods, that is, the foods that not only meet our calorie needs but also have compounds beneficial for our health. Mushrooms fit into this category very well as we will see when we read about why to eat mushrooms.

Mushrooms are not only a quality food but also a way of utilizing agricultural wastes and generating wealth from the waste. The material left after growing mushrooms, commonly referred as spent mushroom substrate (SMS), can be processed into manure. Thus, mushroom cultivation is an important method to promote sustainable manure based farming. Addition of spent mushroom substrate is also reported to improve the soil health. The very fact that mushrooms can be cultivated on paddy straw and number of other agricultural wastes, many of which are just burnt, is sufficient reason to grow mushrooms. By growing mushrooms, we are not only producing a quality food but are also creating a healthy environment. More importantly, it leads to employment generation and women empowerment. With increasing population the land is shrinking and mushroom cultivation utilizes vertical space and requires minimal land making it possible to promote mushrooms in peri-urban and urban areas also. Mushrooms are considered to be the highest protein producers per unit area per unit time.

Why to Eat Mushrooms?

Our ancestors were collecting mushrooms from the forests and fields. These were considered as a delicacy and in some civilizations these were considered as the food for the kings. After cultivation of mushrooms, these have become available to all in large amounts. Now we know that these are a quality food having health benefits. In fact mushrooms are among the best vegetarian food available.

Mushrooms are a rich protein source having essential amino-acids with high digestibility. Mushrooms have all the nine essential amino acids required by the human beings. We have problem of malnutrition in our country. Mushrooms can contribute in solving this problem.

Mushrooms are good for heart as they have low fat and some of the mushrooms have compounds (like lovastatin in oyster mushroom) that are known to lower the cholesterol in the blood. Moreover, mushrooms have low-sodium and high potassium contents and thus are suitable food for persons suffering from high blood pressure.

Mushrooms are a low calorie food with no starch and also have number of antioxidants and hence considered delight of diabetics. These are also rich in fibres and hence good for the intestine and digestive system. These are also a very good source of vitamins especially vitamin B complex. Mushrooms are the only vegetarian source of vitamin D. If we expose mushrooms to UV light after harvest for some time or expose to sunlight, the vitamin D content increases many folds.

These also have vitamin B12 which is not available in plants. These are also rich in minerals, which also include copper (heart-protective) and selenium (anti-cancer).

Characteristics	Benefits
High quality protein	Combats malnutrition
Low sodium high potassium	Controls hypertension
No starch, low sugar	Delight of diabetics
No cholesterol	Healthy heart
Rich in fibre	Improves digestion
Only vegetable with Vit. D	Cures Ricketsia
Folic acid, Vit. B and minerals	Improves health
Ascorbic acid and carotenoids	Anti oxidant
Low calorie food	Reduces obesity
Selenium	Anti cancer property

Fig. 5 : Characteristics and benefits of consuming mushroom

Many of the mushrooms are known to have medicinal and anti-viral properties and their consumption activates the immune system of the human body. Compounds from number of mushrooms have found applications in cancer research and numbers of them have been found to reduce the side effects of radio-therapy and chemotherapy. The benefits have been summarized in Fig.5. As mentioned earlier, mushrooms can be easily digested. Now we know that these can meet our nutritional needs and thus are a health food.

About Medicinal Mushrooms

At present about 30 percent of the world trade is related with medicinal mushrooms. In India however, we have only recently started cultivating medicinal mushrooms. Some species like *Cordyceps sinensis* (keera ghaas) are collected from the forest and are sold. Some growers have started cultivating *Cordyceps militaris* in our country. Technique for cultivation of *Ganoderma* (Reishi mushroom) has been standardised but it is not being promoted because it is known to cause diseases of number of tree species. Major focus in our country is still on edible mushrooms.

Some of the medicinal mushrooms on which research is going on in different parts of the world including trials on human beings are: *Lentinus edodes, Grifola frondosa, Schizophyllum commune, Ganoderma lucidum, Trametes versicolor, Inonotus obliqus, Flammulina velutipes, Phellinus linteus, Cordyceps sinensis.*

Possibly most popular medicinal mushroom is *Ganoderma.*

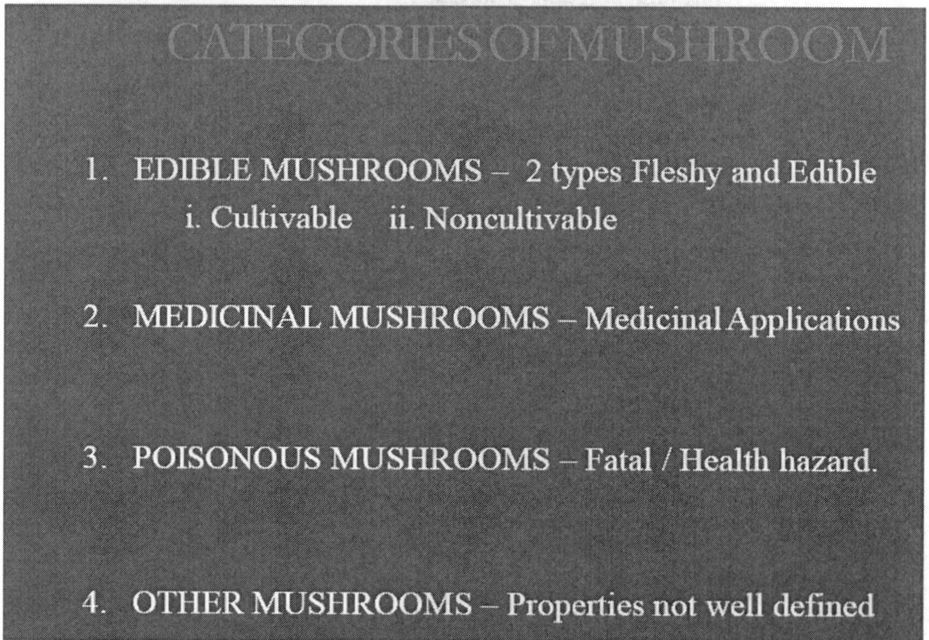

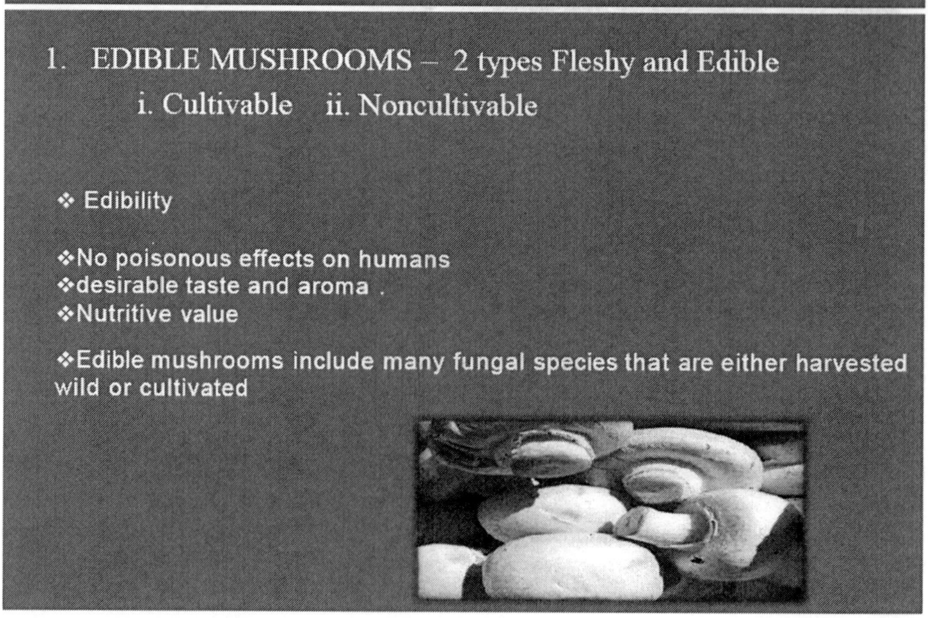

2. POISONOUS MUSHROOMS –

❖Poisonous Mushrooms look like edible mushroom in their morphology and lifecycle. However They can be distinguished by following features:

❖Brightly coloured fruit bodies.

❖Greenish tinge on gills and Yellow-Green spores.

❖Pink coloured spores in gills.

❖Presence of Vulva and Annulus on the Stalk.

❖Oozing of milky or coloured latex at damaged portions

❖Unpleasant odour

•E.g. *Amanita phalloides*

•*Tricholoma muscarium*

1. POISONOUS MUSHROOMS –

***Amanita* sp.**

Lepiota brunneoincarnata

***Cortinarius* sp**

Cortinarius bolaris

***Lepiota* sp.**

Amanita muscaria

1. A.D. 600 - *Aucularia auricula* – 1st mushroom cultivated in China on wood log.

Auricularia auricula

Ganoderma – King of herbs / Mushroom of Immortality

*Letinus edodes (*Shitake) (Both edible & Medicinal) Most popular in Japan

Prevents cerebral hemorrhage Controls Blood pressure level

Chapter 2

Substrates for Growing Mushrooms

Materials for composing a mushroom substrate are diverse and plentiful. Because fungi decompose plant tissue, most homeowners can use by-products generated from gardening, landscaping, tree pruning, and even building projects. Homeowners who collect and pile debris have a perfect opportunity for growing a variety of gourmet and medicinal mushrooms. If a mushroom of choice is not introduced, a wild species from the natural environment will invade. The probability that one of these invading wild mushrooms would be a gourmet species is remote.

Raw Materials

Most by-products from agriculture and forestry industries can make up a base medium for mushroom culture. This base medium is commonly referred to as the "fruiting substrate." This primary material is often supplemented with a carbohydrate- and protein rich additive to enhance yields. Here is a short list of the materials that can be recycled into mushroom production.

- Wood wastes, paper products
- Cereal straws and grain hulls
- Coconut fibers
- Corncobs
- Coffee plants and waste
- Tea leaves
- Sugarcane bagasse
- Banana fronds
- Seed hulls (cottonseed, sunflower, and oil-rich seeds)
- Almond, walnut, pecan, peanut hulls
- Soybean meal, roughage (Okara), and soy waste
- Artichoke waste
- Cactus waste: saguaro and prickly pear, yucca, agave1
- Cereal Straws

For the cultivation of Oyster mushrooms, cereal straws rank as the most usable base material. Wheat, rye, oat, and rice straw perform the best. Of the straws, We prefer wheat. Inexpensive, readily available, preserving well under dry storage conditions, wheat straw admits few competitors. Furthermore, wheat straw has a nearly ideal shaft diameter that selectively favors the filamentous cells of Oyster mushrooms. Chopped into 1- to 4-inch lengths, the wheat straw needs only to be pasteurized by any one of several methods. The approach most easily used by home cultivators is to submerge the chopped straw in hot water (160°F, 71°C) for 1 to 2 hours, drain, and inoculate. First, fill a metal barrel with hot tap water and place a propane burner underneath. (Drums should be food grade quality.

Do not use those that have stored chemicals.) A second method calls for the laying of straw onto a cement slab or plastic sheeting to a depth of no more than 24 inches. The straw is wetted and turned for 2 to 4 days, and then loaded into a highly insulated box or room. Steam is introduced, heating the mass to 160°F (71°C) for 2 to 4 hours.

The semi-selectivity of wheat straw, especially after pasteurization, gives the cultivator a 2-week window of opportunity to establish the gourmet mushroom mycelium. Wheat straw is one of the most forgiving substrates with which to work. Outdoor inoculations of pasteurized wheat straw with grain spawn, even when the inoculations take place in the open air, have a surprisingly high rate of success for home cultivators.

Rye straw is similar to wheat, but coarser. Oat and rice straw are finer than both wheat and rye. The final structure of the substrate depends upon the diameter and the length of each straw shaft. Coarser straws result in a looser substrate, whereas finer straws provide a denser or "closed" substrate. A cubic foot of wetted straw should weigh around 20–25 pounds. Substrates with lower densities tend to perform poorly. The cultivator must design a substrate that allows air exchange all the way to the core. Substrate dynamics are determined by a combination of all these variables.

Paper Products: Newspaper, Cardboard, Books

Using paper products as a substrate base is particularly attractive to those wishing to grow mushrooms where sawdust supplies are limited, such as on islands and in desert communities. Paper products are made of lignin-cellulose fibers from pulped wood, and therefore support most of the wood-decomposing mushrooms described in this book. In recent years, most printing companies have switched to soybean-based inks, reducing or almost eliminating toxic residues. Since many large newspapers are recycling, data on toxin residues is readily available.

Since the use of processed wood fiber may disqualify a grower for state organic certification,Indian cultivators should check with the Organic Certification Director at the State Department of Agriculture before venturing into the commercial cultivation of gourmet mushrooms on paper-based waste products. If these preconditions can be satisfied, the would-be cultivator can tap into an enormous stream of materials suitable for substrate composition.

Corncobs and Cornstalks

Corncobs (sans kernels) and cornstalks are conveniently structured for the rapid permeation of mycelium. Their cell walls and seed cavities provide a uniquely attractive environment for mycelium. Although whole corncobs can be used directly, grinding corncobs to 1–3 inches using hammer-mill type chipper-shredders creates more uniform-sized particles. After moistening, the corn roughage can be cooked for 2 to 4 hours at 160–180°F (71–82°C) to achieve pasteurization. If the kernels are still on the cob, sterilization may be necessary. Cornstalks, having a lower nutritional content, are less likely to become contaminated.

Coffee and Banana Plants

In the subtropical and tropical regions of Central and South America, the abundance of coffee and banana leaves has spurred mycologists to examine their usefulness in growing gourmet mushrooms. The difficulty in selecting any single plant material from warm, humid regions is the speed of natural decomposition due to the activity of competitors. The combination of high humidity and heat accelerates decomposition of everything biodegradable. Leaves must be dried, shredded, and stored in a manner not to encourage composting. Once weed fungi, especially black and green molds, begin to proliferate the suitability of these base materials is jeopardized. Mushrooms demonstrating commercial yield efficiencies on banana and coffee pulp are warm-weather strains of Oyster mushrooms, particularly Pleurotus citrinopileatus, Pleurotus cystidiosus, Pleurotus djamor, Pleurotus ostreatus, and Pleurotus pulmonarius. Reishi mushrooms (Ganoderma lucidum) grow well, and We suspect many other Polypores would also.

Sugarcane Bagasse

Sugarcane bagasse is the major waste product recovered from sugarcane harvesting and processing. Widely used in Hawaii and the Philippines by Oyster growers, sugarcane bagasse needs only pasteurization for cultivating Oyster mushrooms. Some Shiitake strains will produce on sugarcane residue, but yield efficiencies are low compared to wood-based substrates. Since the residual sugar stimulates mycelial growth and is a known trigger to fruiting, sugarcane residues are good complements to wood-based substrates.

Seed Hulls

Seed hulls, particularly cottonseed hulls, are perfect for their particle size and their ability to retain water. Their readily available form reduces the material handling costs. Buffered with 5–7% calcium sulfate and calcium carbonate, cottonseed hulls simply need wetting, pasteurization, and inoculation. Cottonseed hulls, on a dry weight basis, are richer in nitrogen than most cereal straws. Many Button cultivators consider cottonseed hulls a supplement to their manurebased composts. On unamended seed hulls, Oyster and Paddy Straw are the best mushrooms to grow.

Peanut shells have had little or no value except to mushroom growers. Peanut hulls are rich in oils and starch that stimulate mushroom growth. The shells must be chipped into ¼- to ¾-inch pieces, wetted, and pasteurized for several hours. Because peanut shells form subterraneously and are in ground contact, they should be thoroughly washed before pasteurization. Sterilization may be required if pasteurization is insufficient. The addition of 5% gypsum (calcium sulfate) helps keep the substrate loose and aerated. Oyster mushrooms, in particular, thrive on this material.

Soybean Roughage (Okara)

Okara is the main by-product of tofu and tempeh production. Essentially the extracted roughage of boiled soybean mash, Okara is perfectly suited for quick colonization by a wide variety of mushrooms, from the Pleurotus sp. to Ganoderma lucidum, even Morels. Several companies currently use Okara for generating mycelium for extraction and/or for producing flavorings.

All of the above-mentioned materials can be used to construct a base for mushroom production, outdoors or indoors. A more expansive list could include every primary by-product from agricultural or forestry practices. To the imaginative cultivator, the resources seem almost limitless.

Supplements

Supplementing the substrate can boost yields. A wide variety of protein-rich (nitrogenous) materials can be used to enhance the base substrate. Many of these are grains or their derivatives, like rice, wheat or oat brans, ground corn, etc. Many growers use grape pumice from wineries and spent barley from breweries as supplements. Supplementing a substrate, such as straw or sawdust, changes the number and the type of organisms that can be supported. Most of the raw materials used for growing the mushrooms listed in this book favor mushroom mycelium and are nitrogen-poor. Semi-selectivity is lost after nitrogen supplements are added, but ultimately mushroom yields improve. Therefore, when supplements are used, extra care is required to discourage contamination and insure success. Here good hygiene and good flow patterns to, from, and within the growing rooms are crucial.

Supplementation of outdoor beds risks competition from contaminants and insects. If you are supplementing a substrate, the sterilization cycle should be prolonged. Sterilization must be extended from 2 hours for plain sawdust at 15 psi (pounds per square inch) to 4 hours for the same sawdust supplemented with 20% rice bran.

Supplemented sawdust, straw, and compost substrates undergo thermo genesis, a spontaneous temperature increase as the mycelium and other organisms begin growing. If this naturally occurring biological combustion is not held in check, a plethora of molds awaken as the substrate temperature approaches 100°F (38°C).

Below this threshold level, these organisms remain dormant, soon being consumed by the mushroom mycelium. Although true sterilization has not been achieved, full colonization is often successful because the cultivator offsets the upward spiral of temperature. Simply spacing spawn bags or jars apart from one another, and lowering spawn room temperatures as thermogenesis begins, can stop this catalytic climb from starting. For many of the gourmet wood decomposers, a temperature plateau of 75° to 85°F (24° to 29°C) is ideal.

The following supplements can be added at various percentages of total dry mass of the bulk substrate to enhance yields.

- cornmeal
- cottonseed meal or flour
- oat bran, oat meal

- grape pumice
- rice bran
- rye grain
- soybean meal and oil
- spent grains from beer fermentation (barley and wheat)
- vegetable oils
- wheat grain, wheat bran
- nutritional yeast

The nutritional composition of these supplements and hundreds of others are listed . If you are using rice bran as a reference standard, the substitution of other supplements should be added according to their relative protein and nitrogen contents. For instance, rice bran is approximately 12.5% protein and 2% nitrogen. If soybean meal is substituted for rice bran, with its 44% protein and 7% nitrogen content, the cultivator should add roughly ¼ as much to the same supplemented sawdust formula. Until performance is established, the cultivator is better off erring on the conservative side than risking over supplementation.

A steady supply of supplements can be cheaply obtained by recycling bakery waste, especially stale breads. A number of companies transform bakery by-products into a pelletized cattle feed, which also works well as inexpensive substitutes for many of the additives listed above.

Chapter 3

The Mushroom Lifecycle

When a collector finds mushrooms in the wild, the encounter is a mere coincidence, a "snapshot" in time of a far vaster process. The mushroom life cycle remains largely invisible to most mushroom hunters; not so to cultivators. The mushroom cultivator follows the path of the mushroom life cycle. Only at the completion of the mushroom life cycle, which may span weeks or months, do mushrooms appear, and then they occur but for a few days. The stages leading up to their appearance remain fascinating even to the most sagacious mycologists.

For mushrooms to survive in our highly competitive world, where legions of other fungi and bacteria seek common ecological niches, millions of spores are produced per mushroom. A small mushroom, only 4 inches across can produce more than a million spores! With the larger agarics, the numbers become astronomical. Since mushrooms reproduce through spores, the success of the mushroom life cycle depends upon the production of these spore-producing organs.

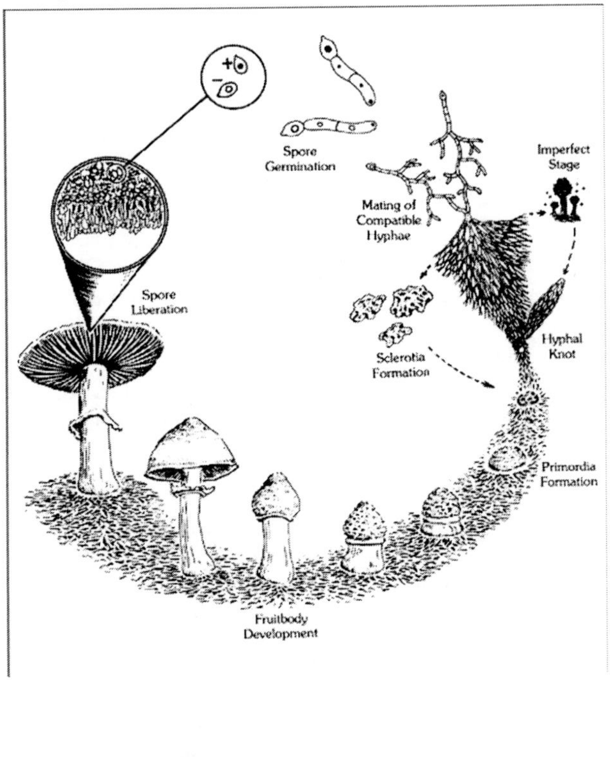

Each spore that is released possesses one-half of the genetic material necessary for the propagation of the species. Upon germination, a filamentous cell called a hypha, extends. Hyphae continue to reproduce mitotically. Two spores, if compatible, come together, fuse, and combine genetic material. The resulting mycelium is then described as being binucleate and dikaryotic. After this union of genetic material, the dikaryotic mycelium accelerates in its growth, again reproducing mitotically. Mated mycelium characteristically grows faster than unmated mycelium arising from single spores.

The Mushroom Life Cycle

The mating of compatible spores is genetically determined. Most of the gourmet species are governed by two incompatibility factors (A and B). As a result, only subsets of spores are able to combine with one another. When spores germinate, several strains are produced. Incompatible strains grow away from each other, establishing their own territorial domains. In this sense, spores from one mushroom can actually compete with one another for the same ecological niche. When compatible spores come together and mate, a much more vigorous form of mycelium fans out.

Each mushroom is like an island. From each center, populations of spores decrease with distance. When spores germinate, the mycelium grows radially, away from the site of origin. Waves of mycelia emanate outwards, and can cover thousands of acres. Overlaying mosaics of mycelium become a complex biological plateau supporting a wide variety of other organisms. Spores, taken up by the wind, or carried by insects and mammals, are dispersed to habitats well distant from the parent mushroom. By coincidence, different varieties of the same species meet and exchange genetic material. In the ever-changing ecological landscape, new varieties are favorably selected for and survive. The driving force of diversity is critical for successful adaptation to environments in flux.

Enzymes and acids are secreted by the mushroom mycelium into the surrounding environment, breaking down lignin-cellulose complexes into simpler compounds. The mushroom mycelium absorbs these reduced organic molecules as nutrients directly through its cell walls. After one mushroom species has run its course, the partially decomposed substrate becomes available to secondary and tertiary saprophytes who reduce it further. Ultimately, a rich soil is created for the benefit of plants and other organisms.

As the mycelium expands, a web of cells is formed, collectively called the mycelial network. The arrangement of these cells is designed to optimally capture an ecological niche. Species differ in the manner by which the mycelial mat is projected. Initially, Morel mycelium throws a thinly articulated mycelial network. (Morel mycelium is the fastest growing of any mushroom I have seen.) Once a substantial territorial domain has been overrun, side branching of the mycelium occurs, resulting in a thickening of the mycelial mat.

With the approach of winter, the mycelial mat retreats to survive in specific sites. At this time, many mushrooms, both gilled and nongilled, produce sclerotia. Sclerotia are a resting phase in the mushroom life cycle. Sclerotia resemble a hardened tuber, wood-like in texture. While in this dormant state, the mushroom species can survive inclement weather conditions like drought, fire, flooding, or other natural catastrophes. In the spring, the sclerotia swell with water and soften. Directly from the sclerotia, mushrooms emerge. Morels are the best-known mushrooms that arise from sclerotia.By the time you find a mature Morel, the sclerotia from which it came will have disappeared.

Most saprophytic mushrooms produce a thick mycelial mat after spore germination. These types of mycelial mats are characterized by many crossovers between the hyphae. When two spores come together and mate, the downstream mycelium produces bridges between the cells, called clamp connections. Clamp connections are especially useful for cultivators who want to determine whether or not they have mated mycelium. Mycelium arising from a single spore lacks clamp connections entirely, and is incapable of producing fertile mushrooms.

While growing through a substrate, the mycelium is growing vegetatively. The vegetative state represents the longest phase in the mushroom life cycle. The substrate will continue to be colonized until physical boundaries prevent further growth or a biological competitor is encountered. When vegetative colonization ceases, the mycelium enters into temporary stasis. Heat and carbon dioxide evolution decline, and nutrients are amassed within the storage vestibules of the cells. This resting period is usually short-lived before entering into the next phase.

Once spores have been discharged, the life cycle has come full circle. Mature mushrooms become a feasting site for small mammals (rodents such as squirrels, mice, etc.), large mammals (deer, elk, bear, humans), insects, gastropods (snails), bacteria, and other fungi. From this onslaught, the mushroom quickly decomposes. In due course, spores not transmitted into the air are carried to new ecological niches via these predators.

Chapter 4

Methods of Mushroom Culture

Sterile tissue culture has revolutionized the biological sciences. For the first time in the history of human evolution, select organisms can be isolated from nature, propagated under sterile conditions in the laboratory, and released back into the environment. Since a competitor-free environment does not exist naturally on this planet, an artificial setting is created—the laboratory—in which select organisms can be grown in mass.

Louis Pasteur (1822–1895) pioneered sterile technique by recognizing that microorganisms are killed by heat, most effectively by steam or boiling water. Tissue culture of one organism—in absence of competitors—became possible for the first time. By the early 1900s growing organisms in pure culture became commonplace. Concurrently, several researchers discovered that mushroom mycelium could be grown under sterile conditions. However, the methods were not always successful. Without benefit of basic equipment, efforts were confounded by high levels of contamination and only after considerable effort was success seen. Nevertheless, methods slowly evolved through trial and error.

The monumental task of creating a sterile environment has been difficult, until recently. The invention of high-efficiency particulate air filters (called HEPA filters) has made sterile tissue culture practicable to all. In vitro ("within glass") propagation of plants, animals, fungi, bacteria, and protozoa became commercially possible. Today, the HEPA (High Efficiency Particulate Air) filter is by far the best filtration system in use, advancing the field of tissue culture more so than any other invention. When air is forcibly pressed through these filters, all contaminants down to .3 microns (μ) are eliminated with a 99.99% efficiency. This means only 1 of every 10,000 particulates exceeding .3 μ pass through. For all practical purposes, a sterile wind is generated downstream from the filter. The cultivator works within this airstream. This unique environment calls for unique techniques. A different set of rules now presides, the violation of which invites disaster.

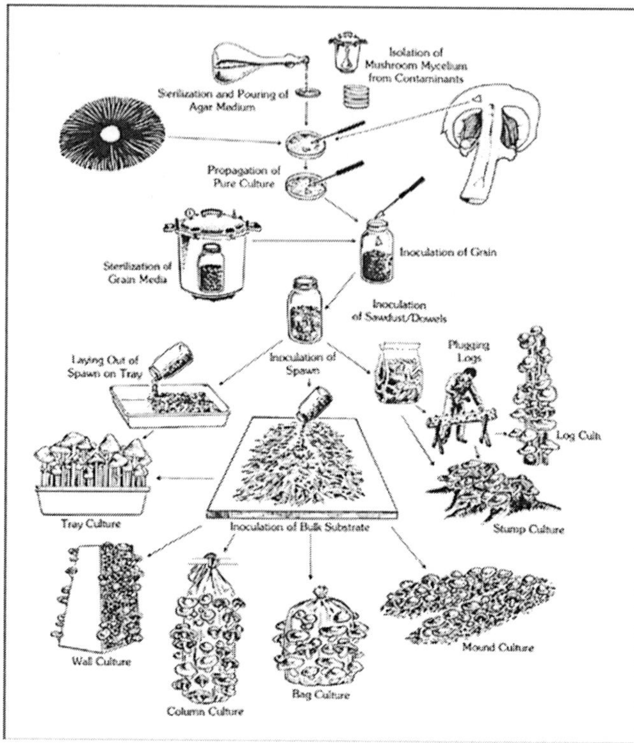

Overview of techniques for growing mushrooms

Sterile tissue culture technique fails if it solely relies on mechanical means. Sterile tissue culture technique is also a philosophy of behavior, ever-adjusting to ever-changing circumstances. Much like a martial artist, the cultivator develops keen senses, a prescience, to constantly evaluate threats to the integrity of the sterile laboratory. These enemies to sterile culture are largely invisible and are embodied within the term "contaminant."

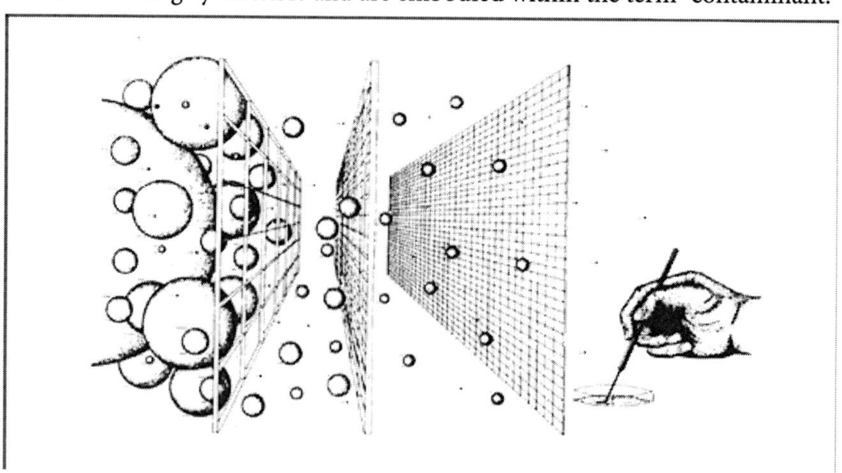

Diagrammatic representation of the effectiveness of filtration media. Dirty air is first filtered through a coarse prefilter (30% @ 10 μ), then an electrostatic filter (99% @ 1μ), and finally through a High Efficiency Particulate Air (HEPA) filter (99.99% @ .3 μ). Only some free-flying endospores of bacteria and viruses pass through.

A contaminant is anything you don't want to grow. Classically, Penicillium molds are contaminants to mushroom culture. However, if you are growing Shiitake mushrooms, and a nearby fruiting of Oyster mushrooms generates spores that come into the laboratory, then the Oyster spores would be considered the "contaminant." So the definition of a contaminant is a functional one—it being any organism you don't want to culture.

The laboratory environment is a sanctuary, a precious space, to be protected from the turmoils of the outside world. Maintaining the cleanliness of a laboratory is less work than having to deal with the aftermath wreaked by contamination. Hence, contaminants, as soon as they appear, should be immediately isolated and carefully removed so neighboring media and cultures are not likewise infected.

Chapter 5

Type of Mushrooms

The Gilled Mushrooms

The gilled mushrooms are the archetypal forms we all know. They are typically umbrella shaped, with a cylindrical stem from which plates radiate outwards from the underside of the cap. Oyster, Shiitake, and Enoki are classic gilled mushrooms. Taxonomically, these mushrooms fall into the order Agaricales. Microscopically, they all reproduce spores by means of club-shaped cells called basidia.

The Himematsutake Mushroom of the Genus Agaricus Agaricus blazei Murrill

Ideal stage for harvesting Himematsutake. Note abundant rhizomorphs at base of stipe. Himematsutake easily pulls from the casing layer and care must be taken to prevent the casing from soiling the white stems.

Description: A classic large Agaricus species, this mushroom is grander in culture than most wild forms. Cap 7–25 cm broad, convex at first, soon hemispheric, then broadly convex, eventually flattening. Often cap margin is smooth, white, and splitting only in age. Cap surface covered with brownish fibrillose patches. Partial veil membranous, floccose with patches of the veil, typically tearing to form a median membranous annulus, but sometimes with remnants attached to the margin at maturity. Gills pallid at first, soon gray, and then chocolate brown when mature. Stem cylindrical, solid, tall, whitish, smooth, flesh thick, often quickly staining yellowish (ochraceous) when bruised. Growing singly or in clusters, arising from stellar sites of dense rhizomorphs leading to and often attached to the stem bases. Imparting strong scent of almonds, especially during cooking. Spore deposit dark chocolate brown en masse.

Fragrance Signature: Musty grain with almond overtones.

Himematsutake growing in Japan on sterilized substrates.

Nutritional Content: On a dry weight basis, protein content can range from 37–48%, making this species one of the most proteinrich of all cultivated mushrooms. Analysis by this author showed, on the basis of a 100 gram sample, 9.88% moisture, 39.3% protein, 1.8% fat, 25.6% fiber, 10.1% ash, and 38.9% carbohydrate. Another un credited analysis posted on the Internet shows that this mushroom has 36.7% protein, 3.4% fat, 6.8% fiber, 7.3% ash, 38.3% sugar, 939 mg/100 g of phosphorus, 18.2 mg/100 g iron, 41.6 mg/100 g calcium, .48 mg/100 g vitamin B1, 2.84 mg/100 g vitamin B2, 345 mg/100 g ergosterol, and 40.9 mg/100 g niacin.

Medicinal Properties: This mushroom produces 1-3 and 1-6 Dfractions of beta glucans, polysaccharides currently under investigation for immunopotentiation. The literature reports beta glucan levels up to 14%. (A recent analysis of the author's Himematsutake showed 9% beta glucans.) Its unique polysaccharides promote natural killer cells that are selectively cytotoxic on tumor cells. This mushroom has been the subject of numerous analyses for isolating constituents, both tumoricidal and immunomodulatory, for the treatment of cancers . The cultured mycelium also produces anti-tumor compounds . That this mushroom produces compounds specifically increasing apoptosis in cancerous cells (but not in healthy cells) and also triggers an immune response, is notable. A yellowish metabolite exuded by the mycelium apparently has bactericidal properties.

Flavor, Preparation, and Cooking: This mushroom imparts a sweet almond flavor, delicate but distinct, a symphony of flavors that linger long after consumption. The aromatics of this mushroom are especially potent in the hours directly after harvest, and the flavor can be potent. When very fresh mushrooms are cooked, the slices of mushrooms undergo a mesmerizing color change from white to strong golden yellow. This color reaction is lost within a day of harvest. The thick stems are especially appealing in their crunchy texture. We recommend cooking them simply on high heat in a bit of olive oil, then seasoning with salt. Cooked mushrooms develop a wonderfully slippery texture, making it one of the most tactilely interesting of all succulent mushrooms. This mushroom is best eaten infrequently, as with daily consumption the novelty of its flavor soon fades. Some people feel the flavor is too strong and prefer the mushrooms be cooked separately so as not to overwhelm the flavors of other dishes. The golden exudate secreted from the spawn also has a very strong almond flavor. After mushrooms are harvested, even at the button stage, the gills continue to mature with spores for days while under refrigeration, becoming pink and eventually dark brown with maturity.Mature specimens rate almost as good as young fruitbodies in their flavor profile.

The Portobello Mushroom of the Genus Agaricus Agaricus brunnescens Peck

Nutritional Content: Crude protein 25–33%; 9% fiber; 10% ash. The FDA has officially designated Button mushrooms as "healthy foods" because they are low in fat and have no cholesterol.

Medicinal Properties: This mushroom contains compounds that inhibit the enzyme aromatase. Aromatase is associated with tumor growth. Compounds inhibiting aromatase have potential for the treatment or prevention of breast cancer . A diet of mushrooms in mice with implanted tumors showed a decrease in aromatase as mushroom consumption increased. However, Agaricus brunnescens contains hydrazines, carcinogenic compounds that have been thought to dissipate only from prolonged, high-temperature heating. More than 80 percent of known hydrazines are carcinogenic. The most notable carcinogenic hydrazine from this mushroom is agaritine, a powerful mutagen, which is activated by the mushroom enzyme tyrosinase, making it heat stable. Enzymes in the digestive system convert agaritine into carcinogenic by-products.

Dutch method of fruiting the White Button variety of Agaricus brunnescens.

The chemical culprits worthy of concern are: 4- (hydroxymethyl)phenylhydrazines and 4 (hydroxymethyl)benzene diazonium ions . Free radicals can also activate Agaricus hydrazines into highly carcinogenic subconstituents as well as catalytic processes in the kidneys . Hence, there are several modes of activating agaritine into highly carcinogenic derivatives.

Flavor, Preparation, and Cooking: Probably the mushroom with the most recipes, this tasty mushroom is stir-fried, cooked into soups, and placed onto pizzas, and is the primary mushroom available in the markets and restaurants of North America and Europe. Button mushrooms have an exceptionally good texture and a much thicker flesh than most. Although it is popular, eating this mushroom raw is not recommended as the nutritional and medicinal benefits are not realized, and there is the danger for negative effects from heat sensitive compounds.

The Black Poplar Mushroom of the Genus Agrocybe Agrocybe aegerita (Brigantini) Singer

Description: A substantial mushroom, often up to 12 inches in diameter. Cap convex to hemispheric, expanding to plane at maturity, smooth, yellowish gray to grayish brown to tan to dingy brown, darker towards the center. Gills gray at first, becoming chocolate brown with spore maturity. Stem white, adorned with a well-developed membranous ring, usually colored brown from spore fall.

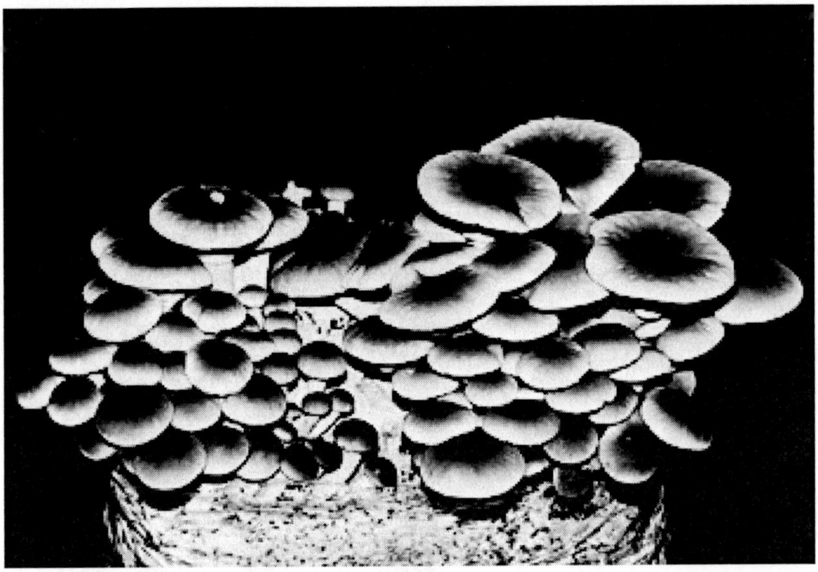

The Black Poplar mushroom (Agrocybe aegerita) fruiting on supplemented aldersawdust/chips 32, 33, 34, and 35 days after inoculation.

Medicinal Properties: None known, although closely related species produce unique antibiotics.

Flavor, Preparation, and Cooking: Finely chopped and stir-fried, cooked in a white sauce and poured onto a fish or chicken, or baked in a stuffing, this species imparts a mild but satisfying, pork-like flavor.

The Shaggy Mane of the Genus Coprinus Coprinus comatus (Muller: Fries) S. F. Gray

Description: Cap 4–10 (15) cm high by 3–4 (5) cm thick, vertically oblong, dingy brown at first, soon white, and decorated with ascending scales. Gills crowded, white to pale, long, broad and slightly attached or free to the stem. Stem 6–12 (15) cm long by 1–2 cm thick, equal, hollow, and bulbous at the base, and adorned with a movable, membranous collar-like ring, that separates from the cap margin as the mushrooms enlarge. As the mushrooms mature, the gills blacken, or "deliquesce," transforming into a black, sporeladened fluid, which drips from the rapidly receding cap margin. The cap eventually totally recedes, leaving only the stem.

Shaggy Manes fruiting from cased horse manure-enriched straw.

Description: Cap 4–10 (15) cm high by 3–4 (5) cm thick, vertically oblong, dingy brown at first, soon white, and decorated with ascending scales. Gills crowded, white to pale, long, broad and slightly attached or free to the stem. Stem 6–12 (15) cm long by 1–2 cm thick, equal, hollow, and bulbous at the base, and adorned with a movable, membranous collar-like ring, that separates from the cap margin as the mushrooms enlarge. As the mushrooms mature, the gills blacken, or "deliquesce," transforming into a black, sporeladened fluid, which drips from the rapidly receding cap margin. The cap eventually totally recedes, leaving only the stem.

Nutritional Content: 25–29% protein (N × 4.38); 3% fat; 59% carbohydrates; 3–7% fiber; and 1.18% ash .

Medicinal Properties: A novel antibiotic has been isolated from this species and is currently being characterized by American researchers. Yingreports that the "inhibition rates against Sarcoma 180 and Ehrlich carcinoma are 100% and 90% respectively." This author knows no other research on the antitumor properties of this mushroom.

Flavor, Preparation, and Cooking: Shaggy Manes were the first mushrooms that seduced me into the art of mycophagy. It may seem odd, but I prefer this mushroom for breakfast. I like to prepare the mushrooms by frying thinly cut dials (stem included) in a frying pan with onions and light oil. Once they are slightly browned, the mushroom can be added to an omelet. Or, fry the mushrooms in butter at medium heat and serve on whole-wheat toast. Many of the recipes listed in this book can incorporate Shaggy Manes. Since this mushroom has considerably more moisture than Shiitake, for instance, the water can be used for steaming vegetables or other foods.

The Enoki Mushroom of the Genus Flammulina Flammulina velutipes (Curtis ex Fries) Singer

Bottle culture of Enokitake in Japan.

Description: Cap 1–5 cm in diameter, convex to plane to upturned in age, smooth, viscid when wet, bright to dull yellowish to yellowish brown to orangish brown. Gills white to yellow, attached to the stem. Stem usually short 1–3 inches, yellow to yellowish brown, darkening with age and covered with a dense coat of velvety fine brown hairs near the base. In culture, the morphology of this mushroom is highly mutable, being extremely sensitive to carbon dioxide and light levels. Cultivated specimens usually have long yellowish stems, small white to yellowish caps. When spores mature, the caps darken to brown.

Nutritional Content: Variable, influenced by substrate components. Crude protein 17–31%; fat 1.9–5.8%; fiber 3.7%; ash 7.4%.

Medicinal Properties: The water-soluble polysaccharide flammulin, is 80–100% effective against Sarcoma 180 and Ehrlich carcinoma . An epidemiological study in Japan found a community of Enoki growers near the city of Nagano that had unusually low cancer rates. Frequent Enoki consumption was thought to be the cause. Ikekawa et al.reported anticancer activity from extracts of this mushroom. Zeng et al. have published studies of the anti-tumor properties of this mushroom. The polysaccharide thought to be active is commonly referred to as FVP for "Flammulina velutipes polysaccharide."

Flavor, Preparation, and Cooking: This mushroom is surprisingly flavorful, including the stems, an opinion not shared by Singer . Traditionally, Enoki is lightly cooked, served in soups or in stir-fries with vegetables, fish, and chicken. The stems are often left long, thus posing some interesting problems in swallowing. I prefer Enoki finely cut, almost diced, and then cooked at high heat for a short period of time. At a recent mycological society gathering, the addition of finely chopped Enoki to a cream sauce, stems and all, resulted in a crème superieur.

The Clustered Woodlovers of the Genus Hypholoma

The Brown-Gilled Woodlover Hypholoma capnoides (Fries) Quelet

The Clustered Woodlover, Hypholoma capnoides, growing from a Douglas fir stump.

Description: Cap orange to orangish yellow to orangish brown to dull brown, 2–7 cm broad at maturity. Convex with an incurved margin, soon expanding to broadly convex to almost flattened, occasionally possessing an obtuse umbo. Cap margin often adorned with fine remnants of the partial veil, soon disappearing, pale yellowish, becoming buff yellow in age. Surface smooth, moist, and lacking a separable gelatinous skin (pellicle). Gills attached, soon seceding, close, white at first, soon grayish, and eventually smokey grayish purple brown in age. Stem 5–9 cm long, enlarged at the base, covered with fine hairs. Partial veil cortinate, sometimes leaving a faint annular zone, becoming dusted purple brown with spores on the upper regions of the stem. Usually growing in clusters.

Medicinal Properties: Given this species' woodland habitat and success in combating competitors, I think H. capnoides should be carefully examined for its antibacterial and medicinal properties.

Flavor, Preparation, and Cooking: Nutty and excellent in stirfries.

H. capnoides fruiting from sterilized hardwood sawdust.

Kuritake (The Chestnut Mushroom) Hypholoma sublateritium (Fries) Quelet

Description: Cap 2–7 cm broad, hemispheric to convex, expanding with age to broadly convex, and eventually plane in age. Cap tan to brown to brick red, darker towards the center, and a lighter yellow near the margin. Margin incurved at first and covered with floccose remnants of the partial veil, soon straightening, and eventually uplifted at maturity. Flesh relatively thick, bruising yellowish. Gills close, bluntly attached to the stem, pallid at first, becoming dark purple gray when mature. Stem 5–10 cm long × 5–10 mm thick, solid, and covered with fibrillose veil remnants below the annular zone. This species often forms large clusters.

Kuritake fruiting on supplemented alder sawdust/chips.

Medicinal Properties: Like H. capnoides, this mushroom is worthy of investigation for its potential medicinal properties. The only reference I have found is a short note in Icons of Medicinal Fungi, which states that "the inhibition rates of Hypholoma sublateritium (Fr.) Quel. against sarcoma 180 and Ehrlich carcinoma (sic) is 60% and 70% respectively" . The reference supporting this claim is in Chinese and lacks further elaboration.

Flavor, Preparation, and Cooking: Excellent in stir-fries or baked. Many of the recipes listed in this book can incorporate Kuritake.

Buna-shimeji (The Beech Mushroom) Hypsizygus tessulatus (Bulliard: Fries) Singer

Description: Mushrooms hemispheric to plane, spotted to marbled with "water spots" on the cap, measuring 2–7 cm. Cap margin inrolled to incurved when young. Mushrooms dark tan, becoming gray tawny brown to creamy brown when mature. Gills bluntly attached to the stem, close, firm, and "wax-like." Stem thick, centrally or eccentrically attached, often tapering towards the base, with variable lengths depending on the method of cultivation.

Wild fruiting of H. tessulatus from cleft in a cottonwood tree.

Medicinal Properties: An unpublished report from the National Cancer Institute of Japan showed strong anti-tumor activity. Mice were implanted with Lewis Lung carcinoma and given aqueous extracts of the fresh mushrooms. At a dose equivalent to 1 gram/kilogram of body weight per day, tumors were 100% inhibited, resulting in total regression of the tumors. Control groups confirmed that, in absence of the mushroom extract, tumors were uninhibited in their growth. No studies with human subjects have been reported.

Flavor, Preparation, and Cooking: A superior edible mushroom with a firm, crunchy texture and mildly sweet nutty flavor, this mushroom is highly esteemed in Japan. Versatile and delicious in any meal, H. tessulatus enhances vegetable, meat, or seafood stirfries and can be added to soups or sauces at the last minute of cooking. The flavor dimensions of this mushroom undergo transformations difficult to describe. Buna-shimeji can be incorporated into a broad range of recipes.

Shirotamogitake (The White Elm Mushroom) Hypsizygus ulmarius (Bulliard: Fries) Redhead

Description: Mushrooms hemispheric to plane, sometimes umbellicate, uniformly tan, beige, grayish brown, to gray in color, sometimes with faint streaks, and measuring 4–15 cm. (This mushroom can become quite large.) Cap margin inrolled to incurved when young, expanding with age, even to slightly undulating. Gills decurrent, close, often running down the stem. Stem eccentrically attached, thick, tapering, and curved at the base. Usually found singly, sometimes in groups of two or three, rarely more.

H. ulmarius, Day 35 and 36 after inoculation into a gallon jar containing sterilized, supplemented alder sawdust/chips.

Medicinal Properties: I know of no published studies on the medicinal properties of Hypsizygus ulmarius. Anecdotal reports, unpublished, suggest this mushroom is highly anticarcinogenic. Much of the research has been done in Japan. Mushrooms in this group, from the viewpoint of traditional Chinese medicine, are recommended for treating stomach and intestinal diseases.

Flavor, Preparation, and Cooking: The same as for most Oyster mushrooms.

The Shiitake Mushroom of the Genus Lentinula Lentinula edodes (Berkeley) Pegler

Description: Cap 5–25 cm broad, hemispheric, expanding to convex and eventually plane at maturity. Cap dark brown to nearly black at first, becoming lighter brown in age, or upon drying. Cap margin even to irregular, inrolled at first, then incurved, flattening with maturity and often undulating with age. Gills white, even at first, becoming serrated or irregular with age, white, bruising brown when damaged. Stem fibrous, centrally to eccentrically attached, fibrous, and tough in texture. Flesh bruises brownish.

Shiitake mushrooms growing on alder log.

Shiitake mushrooms growing on oak logs.

Nutritional Content: Protein 13–18%; niacin (mg/100 g): 55; thiamin (mg/100 g): 7.8; riboflavin (mg/100 g): 5.0. Ash: 3.5–6.5%. Fiber: 6–15%. Fat: 2–5%. Vetter (1995) found that the caps had 15.24% protein while stems had 11%.

Medicinal Properties: Lentinan, a water-soluble polysaccharide (ß-1,3 glucan with ß-1,6 and ß-1,3 glucopyranoside branchings) extracted from the mushrooms, is approved as an anticancer drug in Japan. The Japanese researcher Chihara was one of the first to publish on the anticancer properties of Shiitake, stating that lentinan "was found to almost completely regress the solid type tumors of Sarcoma 180 and several kinds of tumors including methylchloranthrene-induced fibrosarcoma in synergic host-tumour system" . The mode of activity appears to be the activation of killer and helper "T" cells.

Flavor, Preparation, and Cooking: Shiitake can be enjoyed in a wide variety of dishes. A traditional Japanese soup recipe calls for slicing the mushrooms and placing them in a preheated chicken broth complemented with chopped green onions. The addition of miso (a vegetarian substitute for chicken) is also delicious. The Shiitake are steeped in this soup broth for a few minutes and served hot. The flavor and fragrance of slightly cooked Shiitake is tart and totally different than the flavor imparted from thorough cooking. Chinese restaurants usually rehydrate Shiitake and simmer them in the broth of stir-fries.

The Nameko Mushroom of the Genus Pholiota Pholiota nameko (T. Ito) S. Ito et Imai in Imai

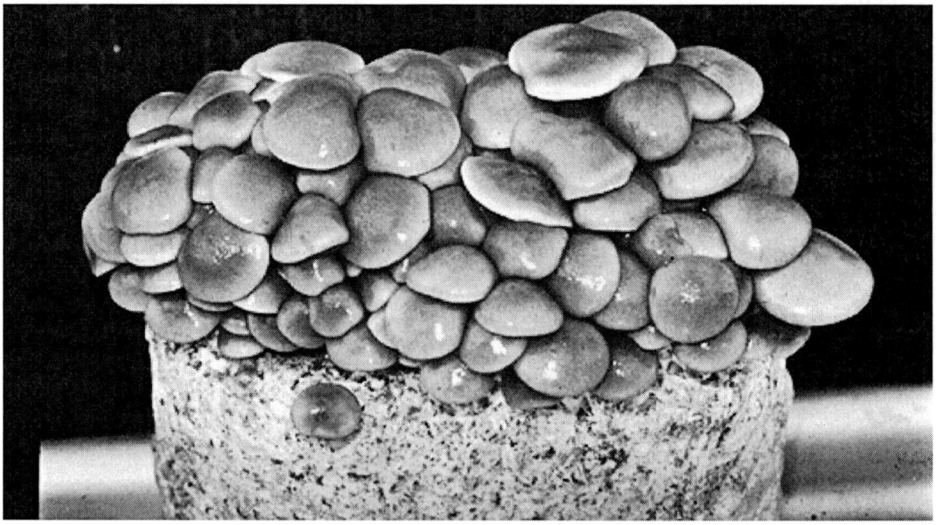

Nameko primordia form 21 days after inoculation on supplemented alder sawdust/ chips.

Description: Cap 3–8 cm, hemispheric to convex, and eventually plane. Surface covered with an orangish, glutinous slime, thickly encapsulating the mushroom primordia, thinning as the mushrooms mature. The slime quickly collapses, leaving a viscid cap. Cap surface smooth. Gills white to yellow, becoming brown with maturity. Partial veil glutinous/membranous, yellowish, adhering to the upper regions of the stem or along the inside peripheral margin. Stem 5–8 cm long, equal, covered with fibrils and swelled near the base.

Nutritional Content: Crude protein (N x 4.38): 20.8%; fat: 4.2%; carbohydrate: 66.7%; ash: 8.3%; and fiber: 6.3%. Vitamins (in milligrams per 100 grams dry weight): thiamin 18.8; riboflavin 14.6; niacin 72.9. Minerals (in milligrams per 100 grams dry weight): calcium 42; potassium 2,083; iron 22.9; sodium 63.

Medicinal Properties: According to Ying , water and sodium hydroxide extracts of this mushroom are 60% and 90% effective, respectively, against Sarcoma 180 implanted in white mice. Furthermore, resistance to infection by Staphylococcus bacteria is substantially improved. This author knows of no parallel studies by Western researchers. The references making these medical claims are in Chinese.

Flavor, Preparation, and Cooking: This is a very slimy mushroom

—a feature that has caused less consternation in Asia than in North America. Nameko is easily diced into miniature cubes and can be used imaginatively in a wide variety of menus, from stir-fries to miso soups. Nameko has a wonderful crunchy texture. Although pleasantly satisfying when lightly cooked, I prefer the strong nutty flavor that thorough cooking evokes. Once the glutinous slime is cooked away, the mushroom becomes very appetizing.

The Golden Oyster Mushroom Pleurotus citrinopileatus Singer

Description: Caps golden to bright yellow, 2–5 cm, convex to plane at maturity, often depressed in the center, thin fleshed, with decurrent gills that show through the partially translucent cap flesh. Stems white, centrally attached to the caps. Usually growing in large clusters arising from a single, joined base. Clusters are often composed of fifty to one hundred or more mushrooms. As strains of this species senesce, the yellow cap color is lost, becoming beige, and fewer mushrooms are produced in each primordial cluster.

A beautiful clustered bouquet of the P. citrinopileatus.

Nutritional Content: Not known to this author. This mushroom is likely to have a similar nutritional profile as P. ostreatus.

Medicinal Properties: According to Yingin Icons of Medicinal Fungi, P. citrinopileatus potentially cures pulmonary emphysema. The supportive references are in Chinese. This mushroom is likely to be similar to P. ostreatus in its cholesterolreducing properties.

Flavor, Preparation, and Cooking: Mushrooms are better broken into small pieces and stir-fried, at high heat for at least 15 to 20 minutes. This mushroom is extremely bitter and tangy when lightly cooked, flavor sensations pleasant to few and disdained by most.

However, when they are crisply cooked, a strong, appealing cashewlike flavor eventually develops. This progression of flavors, primarily affected by the duration of cooking, underscores the Golden Oyster mushroom's versatility as an esculent. This mushroom is becoming increasingly popular as a garnish in salads.

The Abalone Mushroom Pleurotus cystidiosus O. K. Miller

Description: Cap convex to hemispheric, eventually plane, measuring 2–5 cm broad, and cream to dingy white in color. (P. abalonus is reportedly darker colored.) Cap margin often irregular. Gills broad, sometimes widely spaced, strongly decurrent, with irregular edges. Stem thick, centrally or eccentrically attached, and relatively short.

Wild fruiting of P. cystidiosus.

Flavor, Preparation, and Cooking: This mushroom can be cooked like most Oyster mushrooms in stir-fries, white sauces, or adorning lamb.

The Pink Oyster Mushroom Pleurotus djamor (Fries) Boedjin sensu lato

The Pink Oyster mushroom fruiting from bags of sterilized sawdust in Thailand.

Description: Sharing the general shape and appearance of P. ostreatus, except the primordia are bright reddish ("salmon-egg") pink, becoming pinkish as mushrooms develop, eventually a dull pink to light pinkish cinnamon colored, and often fading to straw colored when overmature. (The color transitions are not only agedependent, but vary between strains and are influenced by light conditions.) Cap convex expanding with age to broadly convex to plane. Cap margin inrolled at first, then incurved, and eventually flattening and upturning at maturity. The gills are particularly strongly pigmented with pinkish tones when young, fading to a creamy beige in age. Commonly growing in clusters of multiple mushrooms. When cultivated, variant forms often appear from the same fruiting container as the pink forms. These variants range in color from beige to cream to white, usually with white to gray gills, and often with a highly undulating, scallop-like cap margin.

Nutritional Content: Not fully known to this author. Probably similar to P. pulmonarius. Bano et al. reported that the mushrooms grown from unsupplemented wheat straw, compared to wheat straw supplemented with cottonseed powder, had protein contents of 30% and 36%, respectively. Supplements and spawning rates have a direct impact on the protein content of the mushrooms grown.

Flavor, Preparation, and Cooking: The flavor of this mushroom is not as appealing as many of the other Oyster species listed in this book. Many strains are tougher fleshed and more tart than other Oyster species. The pink color soon disappears upon contact with heat. Upon drying, a majority of (but not all) specimens lose their pinkish tones. Although this mushroom is not my personal favorite, some of my students prefer it to P. ostreatus and P. pulmonarius.

The King Oyster Mushroom Pleurotus eryngii (De Candolle ex Fries) Quelet sensu lato

Description: Cap 3–12 cm in diameter, at first convex, expanding with age, becoming funnel-shaped, with the margin typically in rolled, extending with age. Stem 3–10 cm in length, central, thick, tapering downwards. Gills fairly distant, thin, grayish, and de current. Growing individually or in small groups. Cultivated mushrooms achieve a greater stature and overall size compared to ones collected in the wild.

P. eryngii fruiting from supplemented alder sawdust/chips.

Nutritional Content: Not known, although expected to be similar or exceeding P. ostreatus.

Medicinal Properties: Not known.

Flavor, Preparation, and Cooking: Stir-frying until edges become crispy golden brown. A chewy, nutty mushroom, this species is far superior to P. ostreatus and P. pulmonarius. This mushroom, like other Oyster mushrooms, goes well with Italian dishes, and especially with lamb, pork, and fish.

The Tarragon Oyster Mushroom Pleurotus euosmus (Berkeley apud Hussey) Saccardo

Description: Cap 5–15 cm broad, convex at first, soon broadly convex, expanding to plane, and typically deeply depressed in the center. Mushrooms beige-tan at first, becoming dingy brown with time, sometimes with a hint of blue, becoming light beige tawny in age.

Margin even at first, often irregular in age. Gills dingy, de current, broad, running deeply down the stem. Stem short or sometimes absent.

Nutritional Content: 25% crude protein, 59% carbohydrates, 12% fiber, 9% ash, and 1.1% fat.

Medicinal Properties: Not known to this author. Probably similar to P. ostreatus.

Flavor, Preparation, and Cooking: Versatile and flavorful, this mushroom can be incorporated into a wide variety of recipes. We prefer to sauté young mushrooms at high heat in light oil and to add cashews or almonds along with onions to adorn white fish or salmon that is then baked.

P. euosmus fruiting from pasteurized wheat straw.

The Tree Oyster Mushroom Pleurotus ostreatus (Jacquin ex Fries) Kummer

Description: Cap convex at first, expanding to broadly convex, eventually flat and even upturned in age; 5–20 cm (+) in diameter. White to yellow to grayish yellow to tan, rarely with pinkish tones, to lilac gray to gray-brown. Cap margin smooth to undulating like an oyster shell. Color varies according to the strain, lighting, and temperature conditions. Stems are typically eccentrically attached to the cap. Flesh generally thin. Some strains form clusters; others form individuals.

A sporeless strain of P. ostreatus. Under the microscope, the gill planes are entirely free of a hymenium and basidia.

Nutritional Content: By dry weight: crude protein: 10–30%; vitamin C: 30–144 mg/100 g; niacin: 109 mg/100 g; folic acid: 65 mg/100 g. High in potassium: 306 mg/100 g.

Medicinal Properties: Recent studies show that Pleurotus ostreatus and other closely related species naturally produce Lovastatin (3- hydroxy-3-methylglutaryl-coenzyme A reductase), a drug approved by the FDA in 1987 for treating excessive blood cholesterol. More Lovastatin is present in the caps than in the stems, more concentrated on the mature gills, and especially in the spores. One model showed that plasma cholesterol turnover was significantly enhanced by 50% with a corresponding 25% decrease in the liver compared to the controls . This compound and others related to it may explain the often-reported cholesterolreducing effects of many woodland mushrooms.

Flavor, Preparation, and Cooking: Stir-fry in a light oil at high heat until golden brown and then cook with other condiments.

The Phoenix or Indian Oyster Mushroom Pleurotus pulmonarius (Fries) Quelet, "P. sajor-caju"

P. pulmonarius fruiting from jar of sterilized sawdust.

Description: Cap convex at first, expanding to broadly convex, eventually flat or upturned and often wavy in age; 5–20+ cm in diameter. Grayish white to beige to lilac gray to gray-brown, sometimes with pinkish or orangish tones. (At high temperatures, the cap is lighter in color. Under the same light conditions, under cold conditions, the cap becomes very dark gray to gray black.) Cap margin smooth to undulating like an Oyster. Color varies according to the strain, lighting, and temperature conditions. Stems are typically eccentrically attached to the cap. Veil absent. Flesh generally thin. Strains of this mushroom rarely form clusters of more than 5 or 6 mushrooms.

Nutritional Content: Crude protein (N × 4.38): 14–27%; fat: 2%; carbohydrates: 51% (on a dry weight basis). The variation in the reported protein composition of P. pulmonarius and its close relatives is discussed by Rai et al. (1988). For additional information on the nutritional aspects of this mushroom (identified as "P. sajor-caju"), see Bano and Rajarathnam (1982) and El-Kattan et al. (1991). At 25% protein, this mushroom has about half of the protein represented in a hen's egg, and about one-third of most meats.

Medicinal Properties: Not known to this author.

Flavor, Preparation, and Cooking: P. pulmonarius enhances any meal featuring fish, lamb, and pork. Of course, it also is excellent with most vegetarian cuisines. Slicing and/or chopping the mushrooms and adding them into a stir-fry is the most popular method of preparation. Young mushrooms are far superior to adult specimens, in texture and flavor.

The King Tuber Oyster Mushroom Pleurotus tuberregium (Fr.) Singer

Description: Sclerotia are globose to ellipsoid, whitish to dingy beige on sterilized or pasteurized substrates, darkening with age, and dark brown to nearly black when developing in contact with soils. The initial primordia develop from the sclerotia as a spike-like proboscis, narrow at the apex and broad at the base. Once several inches high, a fluted cap emerges, and the stem develops a coating of darkened fibrils. The cap, centrally attached to the stem, is umbilicate, with decurrent gills, and with a margin, which is initially inrolled at first. Stem scabrous with brownish fibrillose remnants. Quickly maturing, this mushroom can achieve considerable size, up to 10 inches (25 cm) in diameter.

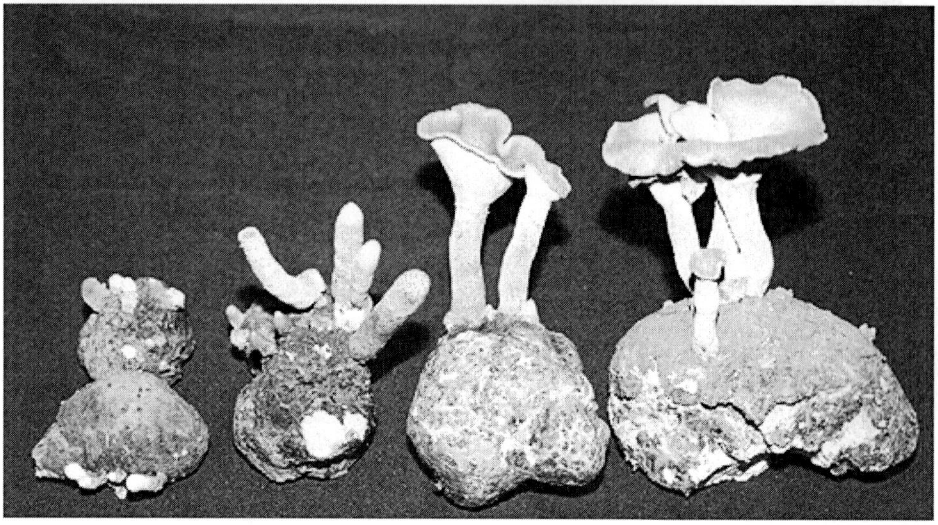

Different stages of mushroom growth from sclerotia.

Nutritional Content: Crude protein (N × 4.38): 14–27%; fat: 2%; carbohydrates: 51% (on a dry weight basis).

Medicinal Properties: Traditionally, this species has been used for treating a wide range of ailments, most prominently topical treatments for skin diseases, and internally for the treatment of heart disease, diabetes, and stomach disorders .

According to Singer , the sclerotia is used by native peoples for such diverse medicinal purposes as stomach pain, constipation, fever, blood pressure, and even smallpox .

Flavor, Preparation, and Cooking: As an edible this mushroom is best cooked at the primordial stage—with the elongated stem and underdeveloped cap. Once the caps begin to enlarge, the texture of this mushroom becomes very tough, and is difficult to consume. Usually the sclerotia are dried first, powdered, and then used as a poultice or an admixture into a soup or tea.

The Caramel Capped Psilocybes (Pacific Coast Teonanacatls: Mushrooms of the Gods) of the Genus Psilocybe

Description: Caps are hemispheric at first, soon convex, expanding to broadly convex and eventually plane in age, 2–10 cm in diameter. Caps are strongly hygrophanous, sometimes chestnut especially when old or when the gills have fully matured. Cap margins are typically even at first, and straightening with age. Some varieties develop a pronounced, distinct, and undulating margin. Gills are colored brown to dark brown, often mottled, and bluntly attached to the stem, typically with a thin whitish margin. The stem is centrally attached to the cap, silky white to dingy brown near the base, often covered with fine fibrils, which may or may not bruise bluish. The stems are cartilaginous, even, straight to sinuous, usually swelling and curved towards the base. The base of the stem is usually fitted with a thick, radiating array of white rhizomorphs.

P. cyanescens complex fruiting on alder wood chips overlaid with a thin layer of straw.

Although mushrooms can be found individually, the majority grows in gregarious or collyboid clusters.

Medicinal Properties: Shamans have used these mushrooms for centuries to diagnose illness. Anecdotal reports received by this author suggest a few venues for possible medical research. An elderly friend suffering from a chronic hearing degeneration told me that small amounts of these mushrooms, too small to cause intoxication, had a remarkably positive effect on his hearing. Residual effects would carry over for several days. He urged me to tell others of his experiences. These mushrooms can enhance the mind's ability for complex visualization, and may possibly be a tool for controlling the autonomic nervous system. They should be used only under the careful supervision and guidance of a physician or shaman. Those unprepared for this experience often suffer from attacks of paranoia, fits of uncontrolled laughter, and other symptoms of psychological alteration.

Western scientists have largely ignored mushrooms of this group for their potential medicinal properties, instead focusing on their short-lived effects on the central nervous system. For a mushroom to manufacture a crystalline structure representing 1% of its mass must be for some evolutionary advantage. Yet the only advantage known is that humans are attracted to them, and thus ensure their survival.

As drugs, the active compounds—psilocybin and psilocin—have virtually no addictive potential for humans. These compounds act as short-lived anti-metabolites and a substitute for serotonin, a primary neurotransmitter in the mammalian brain.

Flavor, Preparation, and Cooking: These mushrooms have an extremely bitter, revolting flavor, and can induce temporary nausea. Thorough cooking may destroy the active compounds in these mushrooms. If boiled for a prolonged period, and the excess water is discarded, the mushrooms are largely denatured of the psychoactive compounds and are edible.

The King Stropharia of the Genus Stropharia Stropharia rugosoannulata Farlow apud Murrill

Description: Cap 4–13 cm reddish brown at first, fading in age, broadly convex to plane at maturity. Margin incurved at first, connected by a thick, membranous veil. Veil breaking with age to form a thick membranous ring radially split with gill-like ridges, usually darkened with spores. Teeth-like veil remnants often seen at the time the ring separates from the cap. Stem thick, equal, enlarging towards the base where thick, white radiating rhizomorphs are attached.

A happy mycophile holding a 3-pound specimen of S. rugosoannulata.)

Flavor, Preparation, and Cooking: Young Buttons, sometimes weighing ½ to 1 pound apiece, can be cut lengthwise to create King Stropharia steaks. Basted with soy sauce and herbal spices, this mushroom is superb on the barbecue during the summer, a time when it produces most prolifically. This mushroom can also be used in stir-fries.

The Paddy Straw Mushroom of the Genus Volvariella Volvariella volvacea (Bulliard: Fries) Singer

Description: As the name implies, this mushroom's most distinctive feature is its volva, or cup, and resembles a classic Amanita except that an annulus is lacking. Cap is 5–15 cm broad, egg-shaped at first, soon expanding to campanulate or broadly convex with a slight umbo. Cap smoky brown to cigar-brown to blackish brown, darker when young, fading in age and/or with exposure to light.

Margin edge radially ridged. Gills free, white at first, soon pinkish, close to crowded. Stem 4–20 cm long × 1.0–1.5 cm thick, white to yellowish, solid, and smooth. The stem base is encased in a thick volva.

Nutritional Content: 26–30% protein, 45–50% carbohydrates, 9– 12% fiber, and 9–13% ash. According to Ying (1987), this mushroom is rich in vitamins C and B, minerals, and assorted amino acids.

Flavor, Preparation, and Cooking: Sliced thin and stir-fried or as a condiment for soups. Used in a wide array of Asian dishes. I like to inject onion-soaked soy (or tamari) sauce via syringe into each Paddy Straw egg, cover with foil, and bake in an oven at 375°F (190°C) for 30 to 45 minutes. This mushroom, when eaten whole, explodes in your mouth creating a flavor sensation par excellence. Canned Paddy Straw mushrooms fail toprovide a flavor experience comparable to fresh V. volvacea.

Renowned mycologist Thaithatgoon with Paddy Straw mushrooms.

The Polypore Mushrooms

Reishi or Ling Chi of the Genus Ganoderma

Ganodermalucidum (Wm. Curtis: Fries) Karsten

Description: Conk-like or kidney-like in shape, this woody-textured mushroom, 5–20 cm in diameter, has a shiny surface that appears lacquered when moist. The cap can be a dull red to reddish brown, and sometimes nearly black in color. Featuring pores on its underside, whitish, browning when touched. Areas of new growth whitish, darkening to yellow brown and eventually reddish brown at maturity, often with zonations of concentric growth patterns. Spores dispersed from the underside collect on the surface of the cap giving it a powdery brown appearance when dry. Stem white to yellow, eventually darkening to brown or black, eccentrically or laterally attached to the cap, usually sinuous, and up to 10 cm in length × .5–5.0 cm thick.

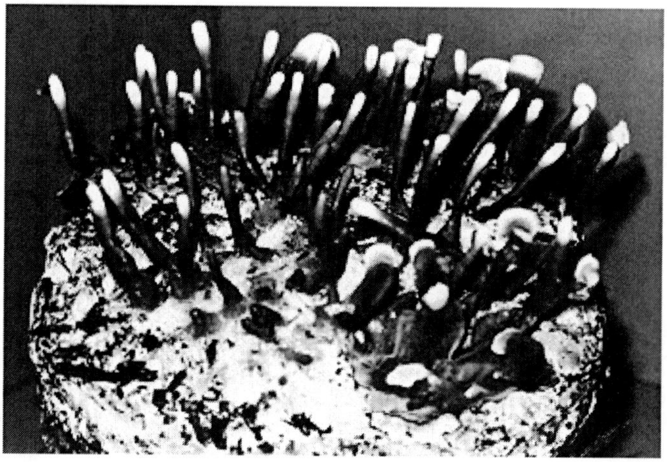

Antler formation of Ganoderma lucidum is controlled by the carbon dioxide of the prevailing environment.

Medicinal Properties: For centuries, the Chinese and Japanese literature has heralded this mushroom for its health-invigorating effects, especially attributing it with increasing longevity, treatment of cancer, and resistance to and recovery from diseases. Himalayan guides have used this mushroom to combat high-altitude sickness.

Mayan Indians traditionally employed Ganoderma in teas to fight a variety of communicable diseases. Reishi has become the natural medicine of choice by many who are afflicted with immunological disorders.

A complex group of polysaccharides, especially the beta-glucans, have been isolated from this mushroom that reportedly stimulate the immune system. (One recent analysis shows our Reishi has 40.6% beta-glucans, an unusually high concentration.) One theory is that these polysaccharides stimulate the production of helper "T" cells, which attack infected cells. Ganoderic acids have also been isolated from Ling Chi and purportedly have anticoagulating effects on the blood and lower cholesterol levels .

Some studies have been published showing its modulating effects on blood pressure and lipid levels , influence on blood glucose levels , and presence of immunomodulating proteins. T. Mizuno summarized the historical development of compounds extracted from this mushroom in Chemical Times . Yun et al.found that in a new mouse model to study lung cancer, Reishi significantly reduced the incidence of tumors. Reishi has also been recommended as a chemopreventive agent against cancer, and markedly reduced the damaging effects of superoxide anions . Mizushina et al.reported two unique DNA polymerase alpha-type inhibitors from Reishi Fruit bodies. Alcohol extracts of Ganoderma lucidum significantly reduced T-4 leukemia cells in vitro. Curiously, these mushrooms produce compounds that are cytotoxic to the growth of cancer cells, while at the same time activating a hostmediated immune response. Studies have shown that, even at large doses, Reishi is nontoxic to healthy cells and is safe to consume. For more information on Reishi's medicinal properties.

Studies by Stavinohaon mice at the Texas Health Science Center in San Antonio showed that unextracted "gill powder" (sic) or ether extracts of the mushrooms showed significant anti-inflammatory activity, comparable to hydrocortisone. Weilnoted that it is difficult to reconcile that a mushroom could be both an immune stimulator and an anti-inflammatory agent. His point is that anti-inflammatory agents generally suppress immune function, not enhance it. Many forms of rheumatoid arthritis are viewed as autoimmune disorders. If the mode of activity of G. lucidum is to act as an immunomodulator, not an immuno-stimulator, then this contradiction in viewpoints is reconcilable.

A novel antiviral compound has been isolated from these mushrooms and shown to reduce replication of HIV. More recently, an extract from this mushroom was found to protect DNA from damage from contact with free radicals and radiation. Wang et al.found that polysaccharides isolated from Reishi potentiate macrophages and lymphocytes, drastically increasing interleukin-1 and interleukin-6.

One of the most thorough descriptions of the active glucans, triterpenoids (ganoderic acids), can be found in Mizuno's publications. Nishitoba et al. found that, as the mycelium matured to form fruitbodies, the triterpene content correspondingly increased, and was more concentrated in older tissue.

Flavor, Preparation, and Cooking: Typically extracted in hot water for teas, tinctures, syrups, and soups. My family enjoys making a tea from fresh, living specimens, breaking them into pieces, boiling in water for 2 hours and then steeping for 30 minutes. The tea is reheated to a desired temperature, strained and served, without sweeteners. If a daily regimen of Ling Chi tea is followed, as little as 3–5 grams per person has been traditionally prescribed. The antler forms, with a reduced hymenial package, have a rich, mildly sweet, and soothing flavor reaction. The welldeveloped conk forms reveal a more bitter aftertaste. Yellow strains are more often bitter than the red and black strains.

Maitake or Hen-of-the-Woods Grifola frondosa (Dicks: Fr.)

S.F.Gray

Wild fruiting of Maitake, G. frondosa, at base of oak tree in a colonial graveyard in upstate New York.

Description: A large, fleshy polypore, dark gray brown when young, becoming lighter gray in age. (Some varieties fade to a light yellow at maturity.) Fruitbody is composed of multiple, overlapping caps, 2–10 cm in diameter, arising from branching stems, eccentrically attached, and sharing a common base. Young fruitbodies are adorned with fine gray fibrils. The pores on the underside of the caps are white.

Nutritional Content: Approximately 27% protein (dry weight). Producing assorted vitamins—Vitamin B1: 1.5 mg%; Vitamin B2: 1.6 mg%; Niacin: 54 mg%; Vitamin C: 63 mg%; and Vitamin D: 410 IU. Saccharide content is nearly 50%. Assorted minerals/metals include magnesium at 67 mg%; iron at 0.5 mg%; calcium at 11.0 mg%; and phosphorus at 425 mg%. Moisture content of fresh specimens are approximately 80%, in contrast to more fleshy mushrooms which average 90% water. Our analysis show beta-glucan levels at 14.5% of dry mass.

Medicinal Properties: In vitro studies at the National Cancer Institute of the powdered fruitbodies (sulfated fraction) of Grifola frondosa show significant activity against the HIV (AIDS) virus when tested through its Anti-HIV Drug Testing System under the Developmental Therapeutics Program. The National Institute of Health of Japan announced similar results in January of 1991. (The mycelium is not active.) Maitake extracts compared favorably with AZT but with no negative side effects. This is the first mushroom confirmed to have anti-HIV activity, in vitro, by both Japanese researchers and U.S. scientists. Studies with human subjects are currently ongoing in United States, Sweden, and England.

Flavor, Preparation, and Cooking: Towards the stem base, the flesh of this mushroom is thick and dense, and is better sliced. The upper petal-like caps are better chopped. This mushroom can be prepared in many ways, delighting the connoisseur mycophagist. Simply slicing and sautéing à la Shiitake is simple and straightforward. This mushroom can also be baked and stuffed with shrimp, sliced almonds, spices, and topped with melted cheese. The late Jim Roberts of Lambert Spawn once fed me this dish featuring a 1-pound specimen that I devoured at one sitting, making for a very satisfying meal and a sleepful night. Dried specimens can be powdered and used to make a refreshing tea.

Zhu Ling or the Umbrella Polypore Polyporus umbellatus Fries

Description: The mushrooms arise from underground sclerotia. The gray to black sclerotium, resembling pigs' dung in form, but woody in texture, swells with water and generates multibranched, circular shaped mushrooms with umbellicate caps. These bouquets of mushrooms arise from a common stem base. The fruitbodies are whitish at first, becoming dingy brown with age, with an underside featuring circular to angular pores. They are often waterlogged due to their high carrying capacity for water.

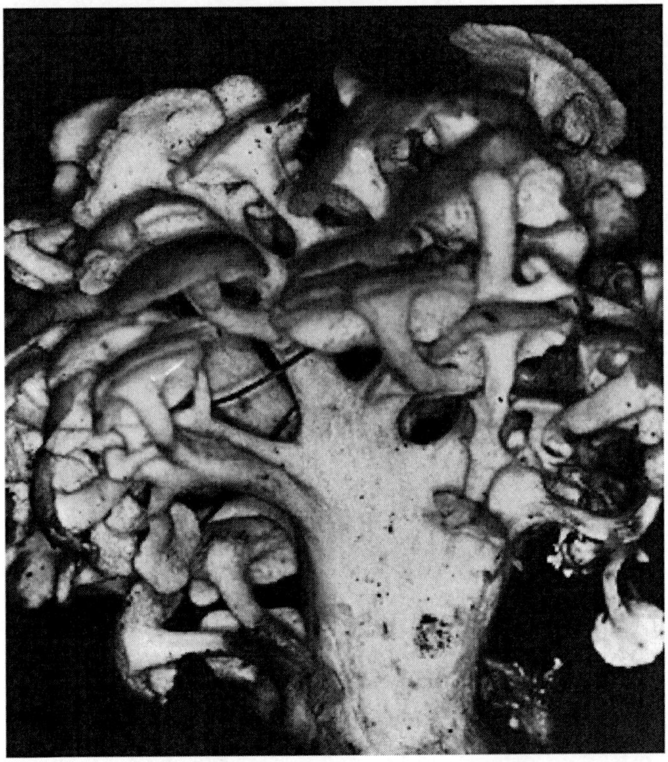

A soft, delicate, fleshy fruitbody arises from the sclerotia.

Nutritional Content: Protein: 8%; coarse fiber: 47%; carbohydrate: .5%; and ash: 7%. The sclerotia, with its woody texture, is likely to have substantially more polysaccharides but less protein than the fleshy fruitbody.

Medicinal Properties: This mushroom has been heralded to possess potent anticancer, immunopotentiating properties. However, few scientific studies have been conducted, and none by Western researchers. During a visit to the Institute of Materia Medica in October of 1983, Dr. Andrew Weil, Dr. Emanuel Salzman, Gary Lincoff, myself, and others were informed by excited researchers that water extracts (tea) of this mushroom, when given to lung cancer patients after radiation therapy, resulted in complete recovery in the majority of the patients. The patients continued a regimen of Zhu Ling tea for several years. Quality of life of the patients dramatically improved, characterized by increased appetites, absence of malaise, and normalization of immune function. The majority of those patients not given Zhu Ling died. I do not know if this study was ever published.

Flavor, Preparation, and Cooking: The fleshy, above-ground mushroom is easily broken or chopped, placed into a frying pan or wok, and sautéed as one would do with Shiitake.

Turkey Tail or Yun Zhi Trametes versicolor (L:Fr.) Pilat

Turkey Tail is a ubiquitous woodland mushroom, growing on diverse hardwoodsand conifers, and is found across much of the globe.

Description: A classic bracket fungus, with multicolored, closely concentric zonations on fan-like shelves, often forming rosettes of overlapping fruitbodies, with slightly hairy surfaces and usually with a wavy outer margin. Mushrooms are highly variable in color, rich with earth tones of grays and browns, often with bluish, greenish, reddish, or whitish zones. Underside covered with whitish pores.

Although an annual mushroom, often surviving for several seasons, the fruitbodies become a dull grayish long after the life cycle has completed. Stem absent or if present, central and short. Often forming great colonies, especially on downed hardwood logs and stumps.

Medicinal Properties: Trametes versicolor is the source of PSK, commercially known as "Krestin," responsible for several hundred million dollars of sales as an approved anticancer drug in Asia. In clinical studies of 224 patientsand 262 patientsafflicted with gastric cancer and treated with chemotherapy, patients showed a decrease in recurrence and an increase in the disease-free survival rate when treatment was combined with a regimen of using the protein-bound polysaccharide (PSK) from Trametes versicolor. By all measures, the treatment was clearly cost-effective. PSK reduces cancer metastasis . PSK stimulated interleukin-1 production in human cell. PSK was also found to be a scavenger of free-radical oxidizing compounds (superoxide anions) through the production of manganese superoxide dismutases . A highly water-soluble, low-cytotoxic polysaccharopeptide (PSP) isolated from this mushroom has been proposed as an antiviral agent inhibiting HIV replication based on an in vitro study . PSP is a classic biological response modifier (BRM), inducing gamma interferon, interleukin-2, and T-cell proliferation, differing chemically from PSK in that it has rhamnose and arbinose while PSK does not but has fucose . Dong et al.reported that a polysaccharide peptide (CVP) and its refined form (RPSP) has not only antitumor properties, but elicits an immunomodulating response by inhibiting the proliferation of human leukemia (HL-60) cells while not affecting the growth of normal human peripheral lymphocytes.

Flavor, Preparation, and Cooking: Tough like leather, the fruitbodies are extracted by boiling in water for use in soups or in teas. This mushroom has been used for centuries as a natural medicine.

The Lion's Mane Mushroom of the Genus Hericium Hericium erinaceus (Bulliard: Fries) Persoon

The teethed fungi have many representatives, some of which are exquisitely edible. Species belonging to the genus Hericium are most notable. Hericium erinaceus and H. coralloides or H. ramosum produce prodigiously in culture and are the best flavored. Hericium abietis, a lover of conifers, is more difficult to cultivate. This group of mushrooms, with their distinctive snow-white icicle-like appearance has long been a favorite of woodspeople. Like Shiitake, the cultivation of this mushroom probably evolved from the astute observations of those who collected them from the wild.

Description: Composed of downward, cascading, non-forking spines, up to 40 cm in diameter in the wild. Typically white until aged and then discoloring to brown or yellow brown, especially at the top.

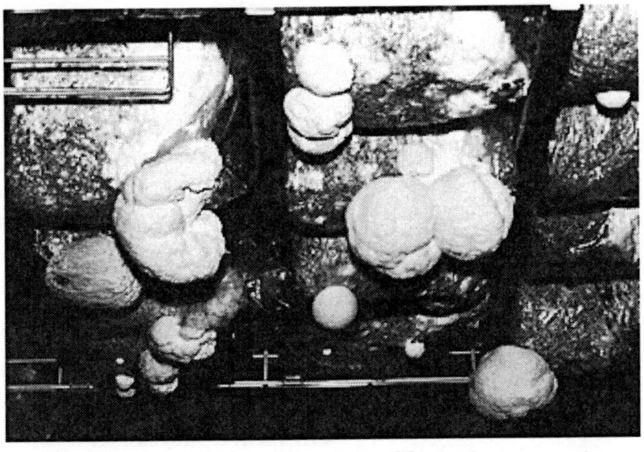

H. erinaceus fruiting from punctured bags in a growing room.

Nutritional Content: Protein 31.7 g/100 g; fat 4 g/100 g; ash 9.8 g/100 g; available carbohydrate 17.6 g/100 g; energy 233 kcal/100 g; dietary fiber 30 g/100 g; sodium 1.2 mg/100 g; phosphorus 1.22 g/100 g; iron 20.3 mg/100 g; calcium 1.3 mg/ 100 g; potassium 4.46 mg/100 g; mangesium 123 mg/100 g; thiamin 5.33 mg/100 g; riboflavin 3.91 mg/100 g; calciferol 240 IU/100 g; niacin 18.3 mg/100 g; ergosterol 381 mg/100 g.

Medicinal Properties: Chenreported that studies on subjects in the Third People's Hospital of Shanghai that H. erinaceus, in tablet form, proved to be effective on "ulcers, inflammations, and tumors of the alimentary canal." Ingestion of this mushroom is saidto have a remarkable effect in extending the life of cancer-ridden patients. Yingreports that pills of this mushroom are used in the treatment of gastric and esophageal carcinoma. I know of no studies outside of China. A patent recently awarded in Japan (#05391544) showed that this mushroom produces Eninacines (sic) (= erinacines), which are strong stimulators to nerve growth factor synthesis . These compounds stimulate neurons to regrow, potentially significant in the treatment of senility, Alzheimer's disease, repairing neurological trauma from strokes, improving muscle/motor response pathways, and cognitive function.

Flavor, Preparation, and Cooking: This is one of my family's favorite gourmet mushrooms. Its flavor is greatly affected by the maturity of the harvested mushrooms, their moisture content, the method of cooking, and particularly the other foods that are cooked with this mushroom. To some, this mushroom has a flavor similar to lobster; to others the flavor is reminiscent of eggplant.

We cut the mushrooms transverse to the spines into dials and cook them at high heat in canola (rapeseed) oil until the moisture has been reduced and the dials are light golden brown. (Garlic, onions, and almonds also go well with this mushroom.) The addition of a small amount of butter near the end of the cooking cycle brings out the lobster flavor. A combination of Shiitake and Lion's Mane, sautéed in this fashion, with a touch of soy or tamari, and added to white rice results in an extraordinarily culinary experience with complex, rich fungal tones.

The Wood Ears of the Genus Auricularia Auricularia polytricha (Montagne) Saccardo

The Wood Ears are peculiar mushrooms that have captured the palate of Asian mycophagists for centuries. Extensive trading of Wood Ears persisted in the late 1800s from New Zealand to China and Hong Kong. Although not remarkably flavorful, these mushrooms rehydrate readily from a dried state, embellishing soups and sauces. Imparting a unique and pleasing texture to most meals, these mushrooms are a centerpiece of Asian cooking, and highly valued. In general, the cultivation methods for this species parallel that of Shiitake on logs or on sterilized sawdust.

Description: A gelatinous cup fungus, ear-shaped, generally purplish grayish brown to dingy brown, 2–15 cm broad, sessile. Covered by a medulla of fine hairs. Surface smooth, wrinkled the center and upturned towards the outer edges. Firmly gelatinous in texture and readily rehydrating in water to natural form.

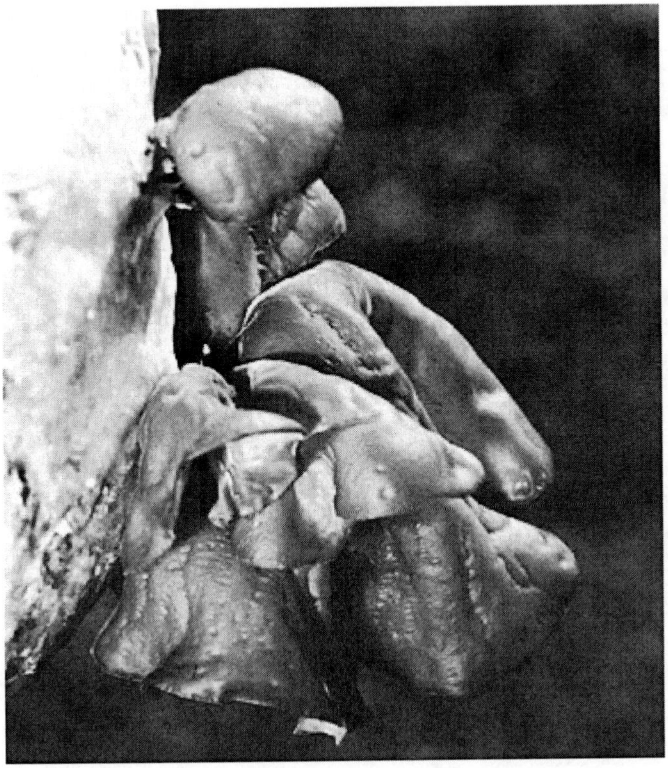

A. polytricha in classic form.

Nutritional Content: 8–10% protein, 0.8–1.2% fat, 84–87% carbohydrates, 9–14% fiber, and 4–7% ash . Moisture content of fresh mushrooms is usually within a few percentage points of 90%.

Medicinal Properties: Yingreports that this mushroom is 8 and 90% effective against Ehrlich carcinoma and Sarcoma 180, respectively. The references are in Chinese. A hematologist at a medical school in Minnesota pricked his finger in a blood-clotting test, and when his blood failed to clot, the ensuing investigation traced the cause to the Wood Ear mushrooms he had eaten the night before at a Chinese Szechwan restaurant. In the United States during 1970s, when Chinese restaurants started serving Wood Ear mushrooms, some patrons noted the emergence of blotchy hemorrhages on their faces the day after consumption. Caucasian women were particularly susceptible. This phenomenon was later dubbed Szechwan Restaurant Syndrome, later becoming known as “Szechwan Purpura” . This discovery lead to a new anticoagulant effective in breaking up blood clots.

Yuan et al.showed that the water-soluble polysaccharide fraction from Wood Ear had a hypoglycemic effect on genetically diabetic mice. A water-insoluble glucan similar to ß 1-3 D-glucans and ß 1-6 D-glucans (560,000–610,00 m.w.) were isolated from the hot-water extract of the fruiting bodies. In vitro tests showed potent anti-tumor activity against the solid form of Sarcoma 180 .

Flavor, Preparation, and Cooking: For many this mushrooms is not remarkably flavorful. Nevertheless, this mushroom adds another dimension to the culinary experience. A. polytricha has a most appealing brittle-gelatinous texture, potentiating the flavors of foods cooked with it. I have seen chefs embellish salads with this mushroom, uncooked, as a garnish.

Wood Ear fruiting from super-pasteurized sawdust.

White Jelly Mushroom of the Genus Tremella Tremella fuciformis Berk

Of the jelly fungi, the Tremella species have drawn the most attention. Existing in its simplest form as a yeast, this group of jellylike mushrooms triggers into a mycelial stage when parasitizing other wood-loving fungi. When wet, they swell with water and are easy to recognize, trembling when you touch them. The bright yellow Witch's Butter, Tremella mesenterica, is probably the most common. Many are cold weather tolerant. They have unique basidia that are vertically partitioned. No poisonous species have been identified, at least so far, and several have had a long tradition of use.

Description: When wet, young fruitbodies swell to resemble blobs of translucent to whitish jelly. When dried the fruitbodies collapse but upon rehydration, the mushrooms reform and regrow, eventually forming ear-shaped leaflets, and as maturing, resembling a chrysanthemum flower, up to 25 cm in diameter.

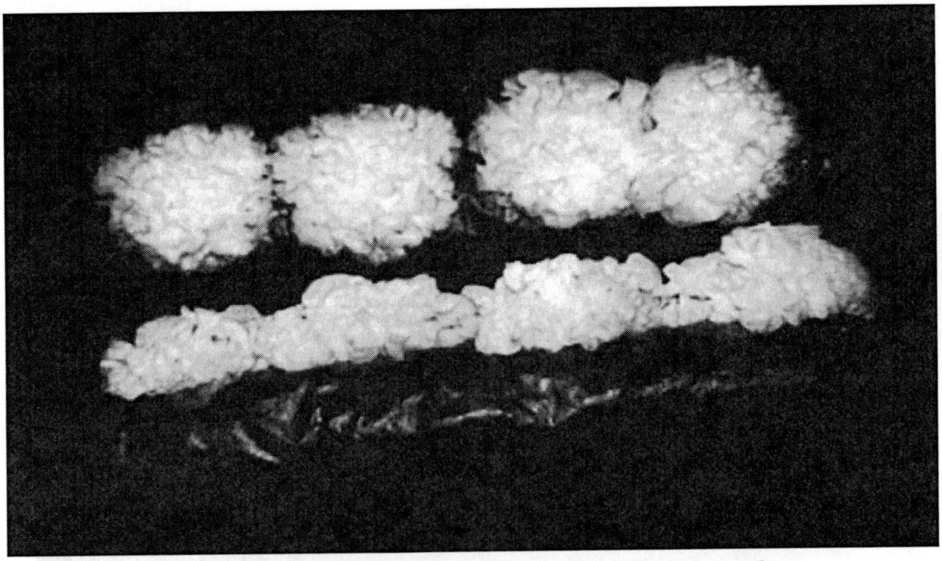

The Silver Ear or White Jelly Mushroom fruiting on sawdust.

Nutritional Content: When dried, mushrooms have 4.6% protein, 0.2% fat, 1.4% fiber, and 0.4% ash.

Medicinal Properties: Tremella fuciformis produces highly watersoluble acidic hetero-polysaccharides, containing xylose, glucuronic acid, mannose, and glucose . Tremella polysaccharides have been studied for their antitumor properties . Gao et al.found that three heteroglycans isolated from the fruitbodies induced human monocytes to produce interleukin-1 and interleukin-6. Extracts from this mushroom are reported to protect canine liver cells from radiation damage , and for the treatment of liver diseases. One of the more complete descriptions of its medicinal properties can be found in Christopher Hobb's book Medicinal Mushrooms, Botanica Press. Many of the references supporting T. fuciformis originate from China, where there has been the longest history of use.

Flavor, Preparation, and Cooking: Mushrooms are usually dried, sealed in plastic, and then sold for addition to soups. They are also made into candies by immersing in water, boiling down, and adding copious amounts of sugar. This mushroom is highly regarded for its gelatinous texture and its primary use as a culinary/medicinal mushroom is in soups.

The Cauliflower Mushroom of the Genus Sparassis Sparassis crispa Wulf ex Fries = Sparassis radicata Weir = ? Sparassis herbstii Peck

A group unto itself, these unique wood lovers have been the favorite of many mushroom hunters who could return to the same location year after year, sometimes finding specimens so huge that they have been difficult to carry! Cauliflower mushrooms belong to the genus Sparassis with five species thus far published: S. crispa, S. laminosa, S. radicata, S. herbstii, and S. spathulata. An eastern form, the same as is found in Europe, with a short stem has been called Sparassis crispa while the western form, sporting a long taproot, has been identified as Sparassis radicata. They may be two forms of the same species.

Description: When young, whitish to yellowish white, looking like a cauliflower, then a brain, tightly-covered egg noodles, soon with ridges elongating into flattened, wavy leaflet-like structures, diverging from the center, smooth, whitish to creamy yellow, with the margins darkening during age or drying. Growing up to 2 ft. (60 cm) across with a central stem, often tapering into a long taproot, darkening towards the base. Cultivated specimens are much smaller than their wild cousins, with mushrooms from the Pacific Northwest old-growth forests providing the largest specimens.

Cauliflower mushrooms fruiting on sterilized hardwood sawdust.

Medicinal Properties: Not known. We would not be surprised if this mushroom contained potent polysaccharides and novel antibiotics. Its ability to grow to a prodigious size and its residence in old growth forests make it a prime candidate for novel constituents.

Flavor, Preparation, and Cooking: A "leafy" fruit body, best torn into pieces and then cooked in mass. Sparassis species are crunchy, with a nutty, mild, but enjoyable flavor. I prefer tearing the leaflets apart, cooking them in a wok, with lots of vegetables, and using the excess water from the mushrooms to provide the steam. The Cauliflower mushroom is exceptionally good when dipped in beerbased batter and deep-fried. Also excellent when added to casseroles, soups, and stews.

The Black Morels Morchella angusticeps Peck complex

Morels embody the mystique of mushrooms. They are elusive, highly camouflaged, and appear for just a few days in any one place. Finding Morels in the wild tests the skills of even the most experienced mushroom hunter. Once found, Morel patches are guarded in secrecy and treasured like family heirlooms. Revealing the location of a Morel patch is an admirable expression of friendship and trust. And it may be the most foolish thing any Morel hunter can do!

The freak occurrence of Morels popping up from nursery-grown, potted phlox plants still mystifies me. This discovery steered me on a fruitless path of experimentation.

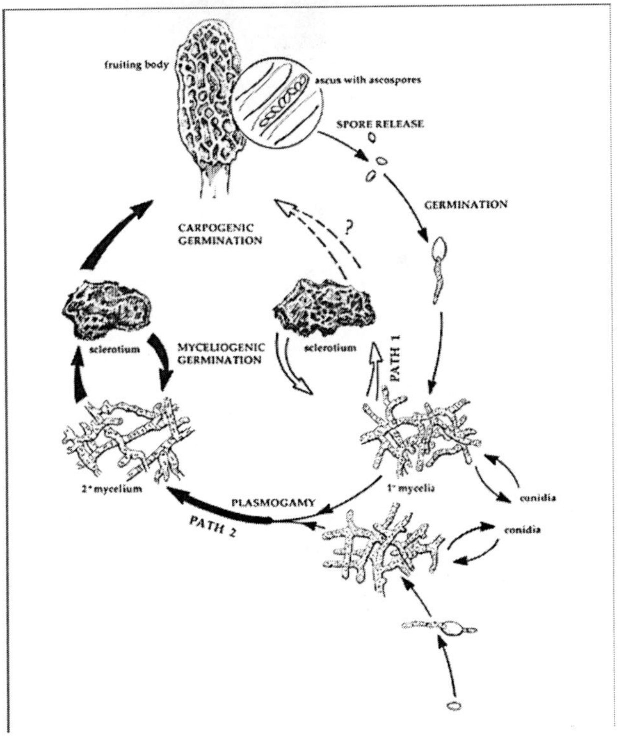

Life Cycle of the Morel by Kandis Elliot,

Morchella esculenta fruiting in trays at Morel Mountain. The technique they have developed is patented.

The unique Morchella atretomentosa.

Description: A honeycombed, ribbed species with black edges, the cap is typically conical shaped, measuring 2–6 cm wide × 2–8 cm high. Stem white, hollow, with a granular texture, measuring 5–12 cm long by 2–4 cm thick. White mycelium is attached to the base of the stem.

Nutritional Content: 20% protein (N × 4.38), 4.8% fat, 8.7% fiber, 64.4% carbohydrates.

Medicinal Properties: Not known to this author.

Flavor, Preparation, and Cooking: A superb edible, this mushroom should be well cooked as many individuals are sensitive to it in the raw state. Morels work well in stir-fries and in a wide range of preparations.

Chapter 6

Cultivation Technology of Paddy Straw Mushroom (*Volvariella Volvacea*)

Introduction

Paddy straw mushroom (Volvariella volvacea), commonly known as the straw mushroom, or the Chinese mushroom, belongs to the family Pluteaceaeof the Basidiomycetes . It is an edible mushroom of tropics and subtropics, and first cultivated in China in 1822. Initially this mushroom was known as "Nanhua mushroom" after the name of Nanhua Temple in Northern Guangdong Province in China. In the beginning, paddy straw mushroom was cultivated by Buddhist monks for their own table, however, by 1875 it was sent as a tribute to the royal family. It is presumed that cultivation of this mushroom begun before the 18th century, almost 300 years ago. Around 1932 to 1935, this mushroom was introduced into the Philippines, Malaysia and other South Asian countries by Chinese.

Paddy straw mushroom is also known as "warm mushroom" as it grows at relatively high temperature. It is a fast growing mushroom and under favourable growing conditions total crop cycle is completed with in 4-5 weeks time. This mushroom can use wide range of cellulosic materials and the C: N ratio needed is 40 to 60, quite high in comparison to other cultivated mushrooms. It can be grown quite quickly and easily on uncomposted substrates such as paddy straw and cotton waste or other cellulosic organic waste materials . It has been considered as one of the easiest mushrooms to cultivate. Paddy straw mushroom was first cultivated in India in 1940, however, its systematic cultivation was first attempted in 1943. Presently this mushroom is more popular in coastal states like Orissa, Andhra Pradesh, Tamil Nadu, Kerala and West Bengal, however, it can also be cultivated in most of the states, where agroclimatic conditions suit and agrowaste is available in plenty.

Biological Characteristics

The fruiting body of the paddy straw mushroom is divided into six different developmental stages viz., pinhead, tiny button, button, egg, elongation and mature stage. Each has its own morphology and anatomy.

1. Pinhead stage: The pinhead stage is of the size of a pinhead in which the veil is spotlessly white. In vertical section, the pileus and stipe are not visible. The whole structure is a knot of hyphal cells.

2. Tiny button: Both the tiny button and pinhead stages are formed from interwoven hyphae. In a young tiny button, only the top of the veil is brown, while the rest is white (Fig. 1). It is round in shape and if a vertical cut is made through the button, the lamellae are seen as a narrow band on the lower surface of pileus.
3. Button stage: This stage of paddy straw mushroom is sold in the market at a premium price. In this stage, the whole structure is wrapped by a coat, which is called as the universal veil (Fig. 1). Inside the veil, closed pileus exists. As such the stipe is not visible but in longitudinal section of mushroom it is visible.

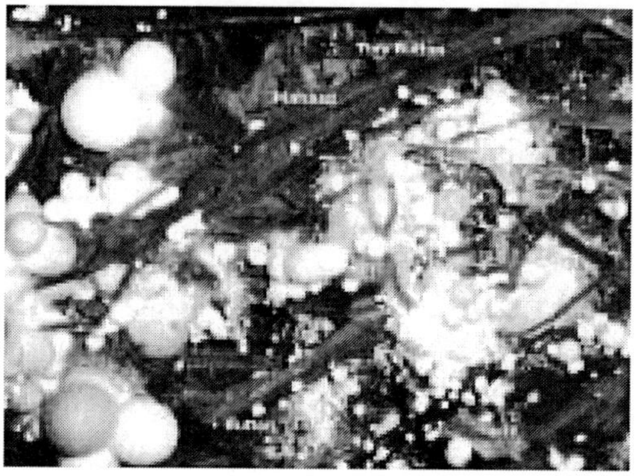

Fig. 1.

4. Egg stage: This stage also fetches premium price in the market, at this stage, the pileus is pushed out of the veil and the veil remains as volva (Fig. 2). The stipe is again not visible at this stage. The lamellae of this stage do not bear basidiospores. The size of the pileus remains very small upto this stage.
5. Elongation stage: The pileus remains close and the size is smaller than mature stage, while the stipe attains the maximum length (Fig. 2). The stipe is marked with water proof drawing ink.

Fig. 2.

6. Mature stage: At mature stage, the structure is divided into three regions: (i) the pileus or cap, (ii) stipe or stalk and (iii) the volva or cup (Fig. 3). The pileus is connected in the centre with stipe and is of usually 6 to 12 cm in diameter.

Fig. 3.

The fully grown pileus is circular in shape with an entire margin and smooth surface. The surface is dark grey at centre and light grey near the margin. The lower surface of the pileus bears lamellae and their number varies from 280 to 380. The lamellae vary in size from full size to one quarter size of pileus. Under the microscope, each lamella is seen to be composed of three layers of interwoven hyphae. The outermost layer is called the hymenium and it forms the clubshaped basidia and the cystedia. The basidia bear basidiospores. Usually one basidium bears four basidiospores. The basidiospores vary in shape; egg shaped, spherical or ellipsoidal. The colour of basidiospores again vary and it may be of light yellow, pink or dark brown.

Another important part of mature fruiting body is the stipe, which connects the volva and the pileus. The length of the stipe depends upon the size of the pileus and it is usually about 3 to 8 cm in length and 0.5 to 1.5 cm in diameter. It is white, fleshy and without any annulus. At the base of stipe remains the volva, which is basically a thin sheet of interwoven hyphae around the bulbous base of the stipe. The volva is fleshy, white and cup shaped with irregular margins. The base of volva bears rhizomorphs, which absorb the nutrition from the substrate.

Nutritive Value

The excellent unique flavour and textural characteristics distinguish this mushroom from other edible mushrooms. The nutritive value of straw mushroom is affected by the method of crop raising and the stages of maturation. The available data reveal that on fresh weight basis, it contains around 90% water, 30-43% crude protein, 1-6% fat, 12-48% carbohydrates, 4-10% crude fibre and 5.13% ash. The fat content increases with the maturation stage and a fully mature fruiting body contains as high as 5% fat. The N-free carbohydrates increase from button stage to the egg stage levels, off at the elongation and drops at the mature stage. The crude fibre remains almost at same level in first three stages and increases at mature stage. The egg stage contains highest level of protein which decreases at mature stage. Ash content remains almost similar at all the developmental stages.

The straw mushroom is known to be rich in minerals such as potassium, sodium and phosphorus. Potassium constitutes the major fraction of major elements followed by sodium and calcium. The levels of K, Ca and Mg remain almost same at different developmental stages except that of Na & P which drop at the elongation and the mature stages. The contents of minor elements, namely Cu, Zn and Fe do not vary much at different stages of development. The levels of thiamin and riboflavin in paddy straw mushroom are lower than Agaricus bisporus and Lentinula edodes, while niacin is at par with these two mushrooms. At all stages, lysine is the most abundant essential amino acid and glutamic acid and aspartic acid are the most abundant nonessential amino acids.

Table 1. Proximate composition of paddy straw mushroom

Content	Composition (quantity/ 100g fresh mushroom)
Moisture	90.40 (g)
Fat	0.25 (g)
Protein	3.90 (g)
Crude fibre	1.87 (g)
Ash	1.10 (g)
Phosphorus	0.10 (g)
Potassium	0.32 (g)
Iron	1.70 (g)
Calcium	5.60 (mg)
Thiamine	0.14 (mg)
Riboflavin	0.61 (mg)
Niacin	2.40 (mg)
Ascorbic acid	18.00 (mg)

Tryptophan and methionine are lowest among essential amino acids. The level of phenylalanine increases nearly one fold at elongation stage, while lysine decreases to about half of its value at the button stage. The straw mushroom is comparable to that of other mushrooms both in terms of amino acid composition and the percentage of essential amino acids in the total amino acids. In fact, paddy straw mushroom contains high percentage of essential amino acids in comparison to other mushrooms and the abundance of lysine is very important. The other three amino acids namely leucine, isoleucine and methionine are low in paddy straw mushroom as compared to other mushrooms.

Table 2. Amino acid contents of paddy straw mushroom

Amino acid	Composition (mg/ 100g protein)
Leucine	3.5
Isoleucine	5.5
Valine	6.8
Tryptophane	1.1
Lysine	4.3
Histidine	2.1
Phenylalanine	4.9
Threonine	4.2
Arginine	4.1
Methionine	0.9

Spawn Production

Spawn is the mycelium of mushrooms growing in its substratum and prepared for the purpose of propagating mushroom production. In a more simple language, it is defined as a medium impregnated with mushroom mycelium that serves as the "seed" for mushroom cultivation. The different stages of spawn production are as follows:

I. Starting Culture

The starting culture can be obtained from any authorized agency or can be raised by any of the following three methods:-

a) Single spore culture Technique

- Selection of unopened mushroom fruiting body, removing dirt with clean cotton followed by wiping of mushroom with 70% alcohol & removing lower portion of stipe with sharp edged knife.
- Placing the fruiting body on pointed end of spiral wire stand placed in a sterilized petridish and covering the whole unit with round mouth beaker.
- Leaving the whole assembly for 30 minutes at room temperature and removing the beaker as well as the fruiting body along with spiral wire stand under a laminar flow chamber followed by covering of petridish aseptically.
- Serial dilution of spore collection upto 10-7 or 10-8 till 10-20 spore/ml count reached and pouring with molten plain agar medium in sterilized petridishes.

- ❖ Incubation of dishes at 32 ± 20C in BOD incubator for 3-4 days and visualization of germlings under inverted microscope for single spore isolates selection.
- ❖ Selection of single spore isolates and multiplying on Malt Extract Agar (MEA) medium by incubating at 32 ± 20C for next 7-10 days in BOD incubator (Fig. 4).

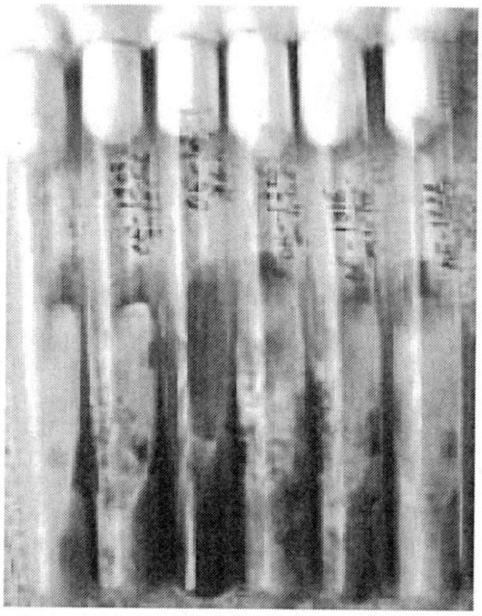

Fig. 4.

b) Multispore Culture Technique

- ❖ The sterilized loop of inoculation needle is used for lifting of the spores from the spore print.
- ❖ The loop bearing thousands of spores is touched on top surface of the petridishes containing the Malt Extract Agar or any other fungal media, followed by incubation of plates at 32 ± 20C for 4-5 days in BOD incubator.

c) Tissue Culture Technique

- ❖ Disinfection of working area and hands with disinfectant and wiping of mushroom fruiting body with 70% alcohol.
- ❖ Making two equal halves of mushroom with the help of sterilized but cooled knife without touching the inner surface of mushroom fruiting body.
- ❖ Removing small pieces of tissue from the stipe pileus connecting point and placing several pieces on the Malt Extract Agar plate on different locations.
- ❖ Incubation of plates at 32 ± 20C for 4 to 5 days in BOD incubator or room temperature (Fig. 5).
- ❖ Transferring of small mycelium bearing portion of medium on the fresh MEA slants followed by incubation at 32 ± 20C for further 4-5 days.
- ❖ Use of cultures directly in spawn substrate.

Fig. 5.

II. Culture Media

There are several media on which the mushroom cultures can grow, the compositions of which are given below:

a) PDA (Potato Dextrose Agar) medium

- ❖ Washing, peeling & slicing of 200g potatoes.
- ❖ Boiling in 1000ml distilled water until potatoes become soft enough to be eaten but not over cooked.
- ❖ Straining through cheese cloth & collecting of liquid in graduated cylinder followed by restoring of volume to 1000 ml by adding fresh distilled water.
- ❖ Addition of 20g dextrose and 15g agar followed by boiling while stirring occasionally until agar is dissolved completely.
- ❖ Transferring of the medium into 10ml tubes or 250ml Erlenmeyer flasks followed by plugging with non-absorbent cotton.
- ❖ Sterilization at 1210C or 15p.s.i for 15-20 minutes.
- ❖ Preparation of slants by putting still hot tubes in slanting position or pouring of medium in sterilized petridishes and leaving as such for next 24 hours for cooling.

b) Malt Extract Agar

- ❖ Water 1000ml, Malt extract 25g, peptone 5g, Agar 20g, pH 7.0 – 7.5.
- ❖ Mixing of weighed quantity of each ingredient in 1000ml of distilled water.
- ❖ Constant heating with intermittent stirring till complete mixing of agar.
- ❖ Pouring of medium in tubes or Erlenmeyer flasks followed by plugging with non-absorbent cotton.
- ❖ Sterilization at 1210C or 15p.s.i for 15-20 minutes.
- ❖ Putting of still hot medium containing tubes in slanting position for slant preparation or pouring of medium in sterilized petridishes followed by cooling at room temperature.

III. Spawn Media

A number of materials, alone or in different combinations are popular as spawn substrates. The most common substrates are rice straw cuttings, sorghum, wheat & rye grains, cotton waste, used tea leaves etc. The protocols adopted for these substrates are mentioned below.

a) Grain spawn (Rye/sorghum/wheat)

About 100 kg grains are first boiled with about 150 liters of water for 20-30 minutes followed by spreading of grains on a sieve for 12-16 hours under shade.

- Mixing of 2kg calcium carbonate and 2kg calcium sulphate with the surface dried cereal grains followed by their thorough mixing and filling in glucose bottles up to 2/3 portion of the available space or in polypropylene (PP) bags of 100 gauge thickness upto their 2/3 available space depending upon the size of the PP bags. Putting of plugs of non-absorbent cotton neither very tight nor very loose.
- Sterilization of glucose bottles or PP bags containing spawn substrate at 1260C or 22 p.s.i for 2 hours followed by cooling under laminar flow bench under aseptic air.
- Inoculation of sterilized spawn substrate with mycelium culture followed by incubation at 32 + 20C for about 2 weeks.
- By the time spawn is ready for use (Fig. 6).

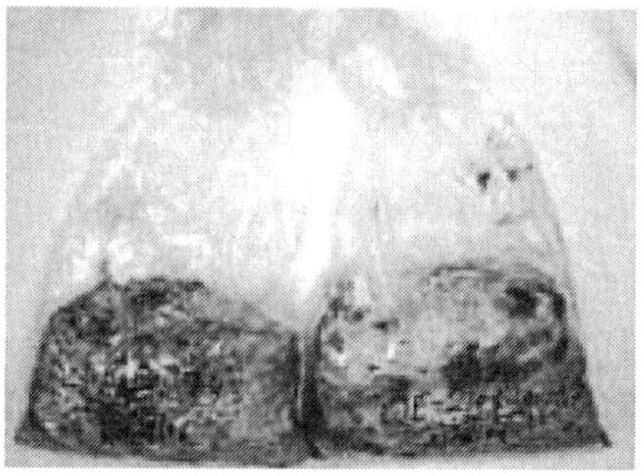

Fig. 6.

b) Straw spawn (paddy straw)

Rice straw is first soaked in water for 2 to 4 hours, then cleaned and cut into pieces of 2.5 to 5cm long. Calcium carbonate and rice bran are mixed @ 1% and 1 to 2%, respectively followed by filling in wide mouth glucose bottles or polypropylene bags of 100 gauge thickness. The bottles/PP bags are closed by plugs of non-absorbent cotton.

- Sterilization of bottles/PP bags containing the spawn substrate at 1260C or 22p.s.i for 2 hours.
- Cooling of spawn substrate and inoculation with mycelia culture under laminar flow chamber.
- Incubation of bottles/PP bags at 32 + 20C for 2 weeks (Fig. 7).

c) Used tea leaves spawn

The used tea leaves are to be first collected & washed to remove any debris, drained & mixed with 2% calcium carbonate for adjusting the pH in the range of 6.8 to 7.8.

- Filling in glucose bottles or polypropylene bags & rest procedure is as for the grain or straw spawn.

Card fly grade of cotton waste is mainly used for spawn making. The further protocol is similar to that for used tea leaves.

e) Manure-husk spawn

A mixture of fresh horse manure and lotus seed husk is used in equal proportions, first by steeping in water until enough moisture is absorbed. The compost is piled upto 1m height in the form of a pyramid and left as such for next 4 to 5 days. The pile is broken and if needed additional quantity of water is added and repiled. This is repeated 5 times after every 4 to 5 days.

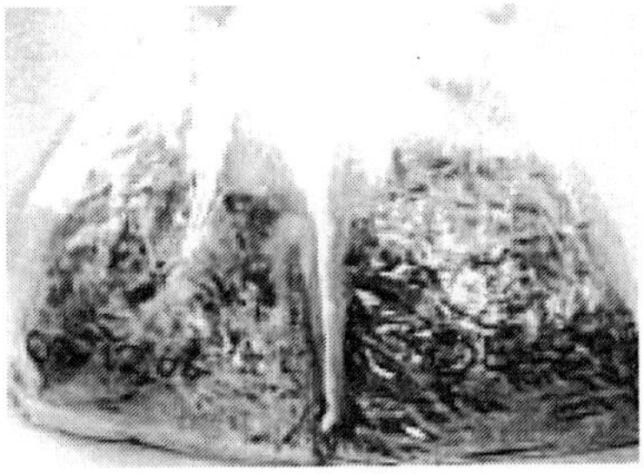

Fig. 7.

- The compost is filled in glucose bottles or in air tight aluminium canes and sterilized.
- After cooling of the compost, mushroom mycelium is inoculated and the compost is incubated at 32 + 20C for 2 weeks or till the spawn is ready.

IV. Key points of Spawn making Protocol

1. Substrates such as rice straw and cotton waste etc. should not be over wet as if water stands on the bottom, mycelium will not grow.

2. Container should not be tightly sealed as air can not escape and steam can not enter properly. Autoclaving will be imperfect.

3. One can prevent entry of moulds from outside after sterilization by:

a) Using only very clean cotton stoppers.

b) Leaving at least 3-4 cm free space between lower surface of cotton stopper and substrate.

c) Avoiding the spoilage of container walls around the stopper and between stopper and the substrate surface.

d) Preventing cotton plug from getting wet during autoclaving by covering loosely with aluminium foil.

e) Keeping the outside of the containers clean as far as the aluminium foil reaches.

4. Inoculate under clean conditions in a room without any air movement (close door and windows) or under laminar flow bench.

a) Clean the table with disinfectant

b) Disinfect the hands with soap and disinfectant

c) Safe transfer of autoclaved substrate in the inoculation room

d) Use only pure culture spawn

e) Cover opening after inoculation with aluminium foil and press around the neck of the container.

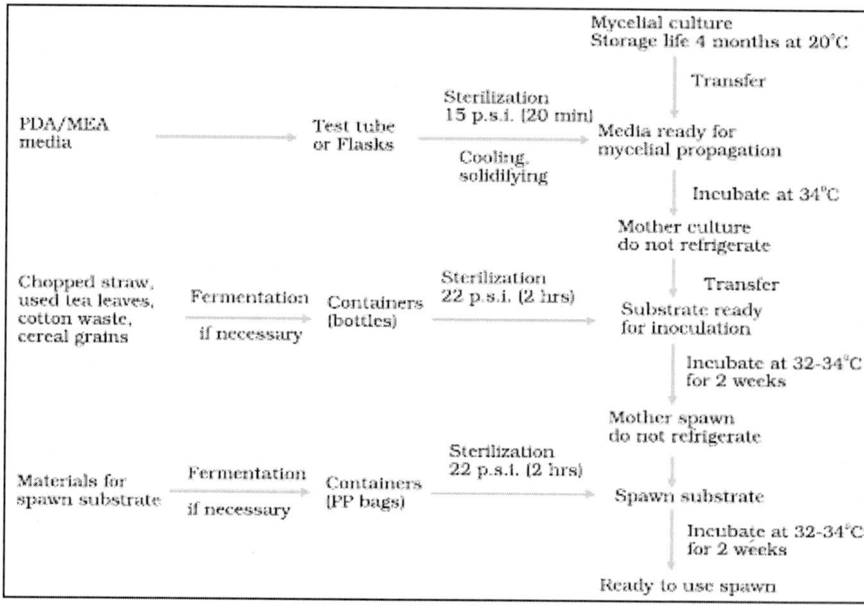

Fig. 8. Spawn production flow chart for V. volvacea

5. Place spawn under optimal growth conditions. Spawn which is not needed for inoculation can be used for mushroom production.

V. Storage of Pure Cultures and Spawn

The optimal temperature for growth of V. volvacea ranges from 30-350C and the most suitable temperature is 320C. Mycelium does not grow at all when the temperature is raised to 450C or dropped to 150C. At a temperature of 15±10C most strains of V. volvacea survive for the longest period .

After complete colonization of spawn substrate with mushroom mycelia, it is ready to be used. However, if it is not to be used immediately, then it should be removed from incubator and stored at a lower temperature to prevent further growth, aging and death. The storage temperature should range from 15-200C as at this temperature the growth of mycelium is arrested and mycelia are unharmed and remain viable for longer period.

Mushroom Cultivation Technology

A variety of waste materials have been used for cultivation of the paddy straw mushroom, which include: paddy straw , water hyacinth , oil palm bunch , oil palm pericarp waste , banana leaves & saw dust , cotton waste , sugarcane bagasseetc. Paddy straw mushroom prefers high cellulose, low lignin containing substrate and produces a family of cellulolytic enzymes.

The cultivation of Volvariella is less sophisticated, less extensive and can be rewarding in tropical & subtropical climates.

Before 1970, it was only paddy straw, which was in use for paddy straw mushroom cultivation. However, in 1971, cotton waste Ginning mill waste) was first introduced as the heating material for growing of straw mushroom followed by complete replacement of paddy straw with cotton waste by 1973 in Hong Kong. This was practically the turning point in the history of paddy straw mushroom because cotton waste gives a higher and more stable yield (30 to 40%) along with early fructification and harvesting. After adoption of cotton waste, the cultivation of paddy straw mushroom has become semi-industrialized in Hong Kong, Taiwan, Indonesia, China & Thailand.

The common methods employed for paddy straw mushroom cultivation are given below:

a) Conventional Method

The different steps involved in this method are as follows:

- ❖ Preparation of paddy straw bundles of 0.75 – 1.0 kg (80-95cm long & 12.16cm wide) preferably from hand threshed paddy.

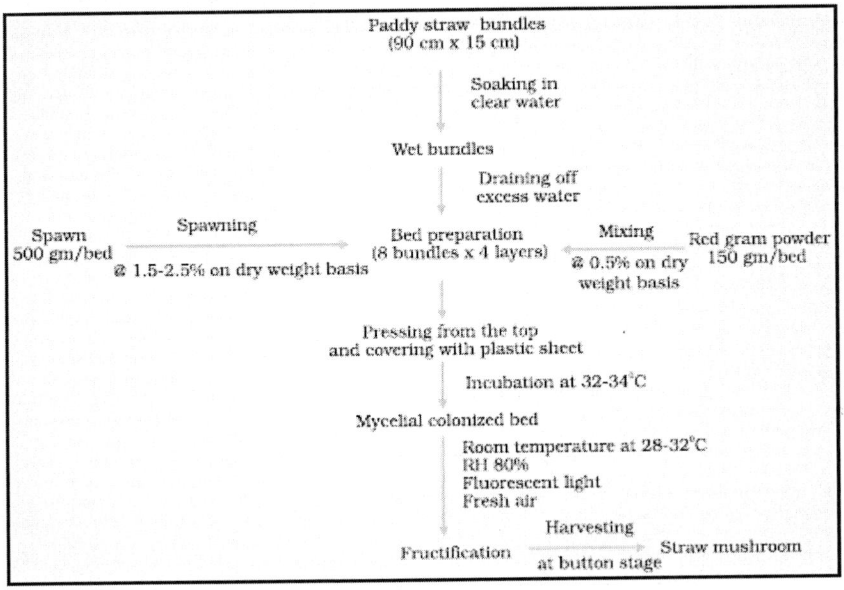

Fig. 9. Conventional method of paddy straw mushroom cultivation

- ❖ Immersing of bundles in clean water for 12-18 hours in a cemented water tank.
- ❖ Draining out of excess water by placing bundles on raised bamboo platform.
- ❖ Making bed by placing 4 bundles side by side and another four bundles similarly but from the opposite side, forming one layer of eight bundles. The open ends of bundles from opposite sides should overlap in the middle.
- ❖ Forming of second, third & fourth layer by intermittent spawning between first and second, second and third and third and fourth layers.

- ❖ Spawning on entire surface of the layers of the beds at a space of 5cm apart leaving margin of 12-15cm from edges.
- ❖ Sprinkling of red gram powder over the spawned surface.
- ❖ Using 500 gm spawn and 150 gm of red gram powder for a bed of 30-40 kg of dried paddy straw.
- ❖ Pressing of bed from the top and covering with clean plastic sheet for maintaining required humidity (80-85%) and temperature (30-350C).
- ❖ Removing of plastic sheet after 7-8 days of spawning and maintaining temperature of 28-320C and relative humidity about 80%.
- ❖ Mushroom will start appearing after 4-5 days of sheet removal and will continue for next 20 days.
- ❖ After crop harvest the substrate can be used for manure in the field.

Note:

- ❖ For hot regions the width of bed can be decreased by placing first layer of 4 bundles followed by another layer of 4 bundles from opposite side but directly on the first layer. It is to be followed in 3rd, 4th & 5th layers. The 5th layer can be of bundles or of loosened paddy straw.
- ❖ The size of beds may vary from 100cm x 100cm x 100cm; 60cm x 60cm x 30cm and 60cm x 60cm x 120cm.
- ❖ Alternatively the beds can be prepared with the help of boxes of 80cm x 80cm x 10cm and of 60cm x 40cm x 30cm size. In this method the material is to be chopped to a uniform length of 20cm and followed by filling in box parallel with the length of the box. It is followed by soaking of the material along with box in 2% CaCO3 solution for 2 hrs or until the straw becomes dark brown. It is followed by draining of excess water and spawning of substrate at a depth of 5 cm from the sides of the box followed by plugging the openings with previously water soaked newsprint. The boxes are to be incubated at a temperature of 35 to 380C with RH of 75% for next 4-5 days.

It is followed by lowering of temperature to 28 to 300C with RH of 75 to 85% along with introduction of fresh air. Use only superfine mist for maintaining proper level of humidity in the room. The bedding material can also be sprayed with fine mist if it is getting dried. Controlling of ventilators is must for maintaining optimum aeration and temperature inside the room.

b) Improved Cage Cultivation

i) Material required

1. Paddy straw 60/Cage bundles
2. Spawn bottle 2/Cage
3. Wooden cage 1 No. (1m x 50cm x 25cm)
4. Drum 1 No. (100 liter capacity)
5. Polythene sheet 4 meters

6. Binding thread 3 meters
7. Sprayer/Rose can 1 No.
8. Dithane Z-78/ 1 Pkt. (200 gm) Bavistin
9. Malathion 1 bottle (250 ml)
10. Dettol/Formalin 1 bottle (1/2 liter)
11. Dao (Hand 1 No. chopper)
12. Thermometer 1 No.

ii) Methodology

- Select dry, fresh and hand threshed paddy straw free from moulds and leafy portion.
- Make 25 cm long and 10 cm thick bundles @ 60 bundles for each cage (Bed).
- Soak the bundles in boiling water for 20-30 minutes and allow cooling and draining of excess water.
- Disinfect the cage and polythene sheet with 2% Formaline or Dettol solution.
- Arrange ten bundles uniformly in the cage as the bottom layer and put some spawn grains over and inside the bundles (Fig. 10). Put a second layer of ten bundles over the first and spawn as before. Repeat this till six layers of bundles are achieved or till filling of the cage.

Fig. 10.

- Spray solutions of 0.1% Malathion and 0.2% Dithane Z-78 all over the bed. Cover the whole bed with polythene sheet and bind securely with a binding thread.
- Keep the spawned cage in a room or a shed for mycelial run (Fig. 11). A warm place with temperature around 300C is helpful for better mycelia growth.

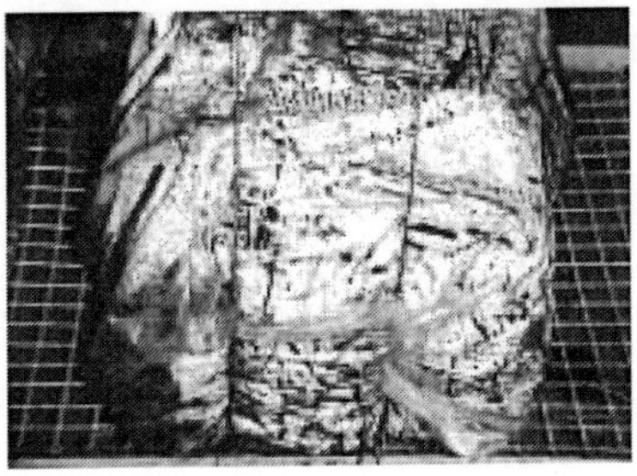

Fig. 11.

- Remove the polythene sheet after the mycelial run is completed (Fig. 12). Maintain high humidity in and around the bed till pinheads appear (Fig. 13).

Fig. 12.

- Pinheads appear within 10-15 days after spawning. Harvest at the egg stage (Fig. 14).

Fig. 13.

Fig. 14.

- Continue spraying water for the next flush of mushroom to appear within a week or so.

c) Outdoor Method

The best place to cultivate paddy straw mushroom outdoor is in shade created by trees or creepers. The steps involved are as follows .

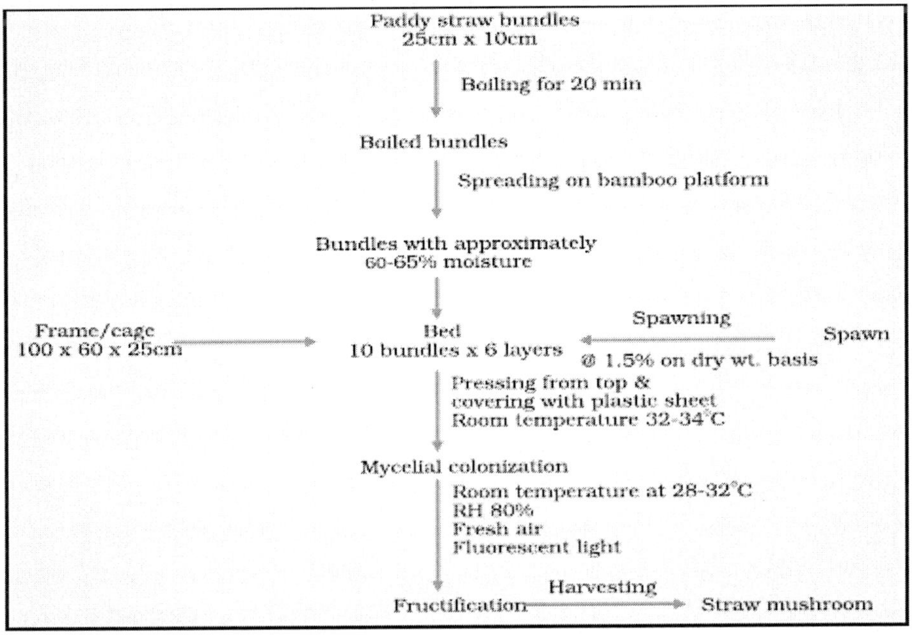

Fig. 15. Cage method of paddy straw mushroom cultivation

- ❖ Preparation of raised platform either with sand or bamboo poles or wooden planks or bricks.
- ❖ Preparation of bundles of 40cm length and 10 cm width.
- ❖ Soaking of bundles in running water or in 2% $CaCO3$ solution.
- ❖ Driving of bamboo pole into the center of each end of the bed.
- ❖ Preparation of layer of bundles followed by spawning.
- ❖ Laying down of 4 layers of bundles during summer months and 7 layers during rainy season.
- ❖ Topping of bed with 20cm deep layer of rice straw followed by covering with polythene sheet.
- ❖ Removing of polythene sheet after 4 days and sprinkling of water carefully on 6th day.

Spraying of water can be avoided during rainy season.

- ❖ Prohibit spraying of water after appearance of the mushroom pinheads.

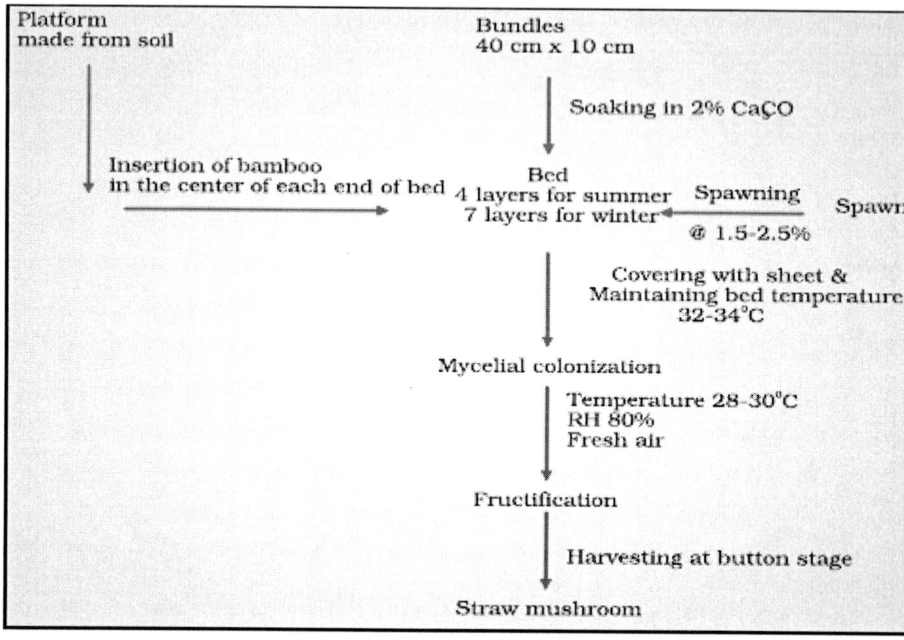

Fig. 16. Outdoor method of straw mushroom cultivation

d) Indoor Method

The indoor method can be divided into following 5 steps :

i) Substrate: Cotton waste is the preferred substrate for cultivation of paddy straw mushroom by this method. However, paddy straw can also be used. Cotton waste is preferred over paddy straw as it contains more cellulose and hemi-cellulose (Table 3). The fine texture of cotton waste helps in retention of moisture, which minimizes the water requirement at later stages of cropping, and thus helps in avoiding damage to mushroom primordia.

ii) Compost preparation:

Substrate is wetted with 1% lime (on dry weight basis) and for cotton waste, a square wooden rack (92x92x28cm) is used for holding a layer of cotton waste about 30cm deep. The workers are used to tread the cotton waste so that it absorbs sufficient quantity of water. After first layer is trodden another layer isapplied. This process is repeated until the required quantity is trodden. However, in case of paddy straw, pile is made (1.5 m high x 1.5 m wide) nby adding sufficient quantity of water mixed with 1% lime (Fig. 18). The pile is also made with wet cotton waste and left to ferment in the open but under cover during rainy season or extreme cold. First turning is given after 2 days and 5% rice bran is mixed in case of paddy straw substrate and water is added if needed.

Table 3. Carbohydrate composition in popular substrates of paddy straw mushroom

Carbohydrate	Cotton waste	Rice straw	Banana leaf
Total Nitrogen	1.22	0.66	1.71
Total Carbon	49.94	54.26	50.52
C:N ratio	40.90	84.00	29.50
Hemi-cellulose	8.73	17.11	19.95
Cellulose	50.76	29.68	10.85
Lignin	10.47	12.17	18.21

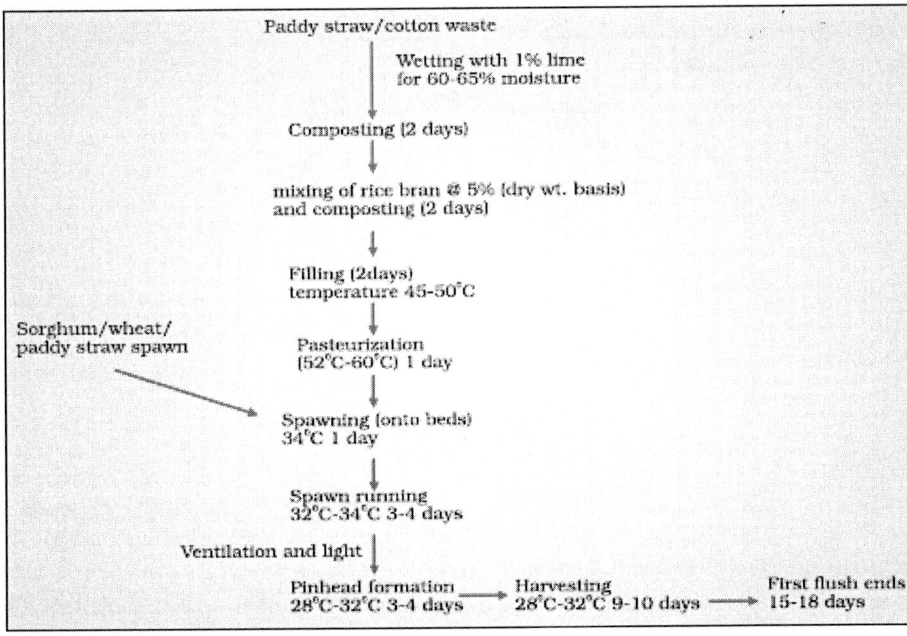

Fig. 17. Indoor method of paddy straw mushroom cultivation

Fig. 18.

However, nothing is added in case of cotton waste substrate. Again pile is formed and left for fermentation for the next 2 days.

iii).Bedding and Pasteurization:

The compost is spread on shelves in rooms or in pasteurization tunnel in suitable thickness (Fig. 19). The thickness of the substrate on shelves varies in different seasons from 5 cm to 10 cm thick. During summer months, lesser thickness is needed, while higher in winter to preserve moisture and heat. The surface is made even by pressing lightly. After 8-12 hours of compost spreading live stream is introduced with the help of rubber hose of 6cm in diameter. A temperature of 620C is maintained for 2 hours for cotton waste compost and of 650C for 6 hrs for paddy straw compost. After steaming, the shed or room is closed to keep a temperature of 500C for next 24-36 hrs and followed by natural cooling of the substrate. The compost is spawned on reaching the temperature near 350C.

Fig. 19.

iv) Spawning: The compost is spawned with fresh spawn @ 1.4% of dry weight or 0.4% of wet weight basis of the compost (Fig. 20). The pieces of broken spawn are inserted in compost at a depth of 2 to 2.5 cm at a distance of 12 to 15 cm. The spawn is covered with displaced compost and the bed is covered with thin plastic sheet. The room temperature is maintained at 32 to 340C during spawn running and the compost will be colonized within next 4-5 days in cotton waste and 5-6 days in paddy straw compost (Fig. 21).

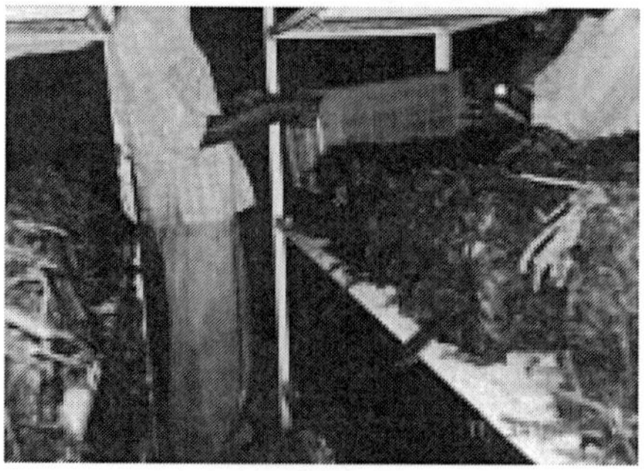

Fig.20

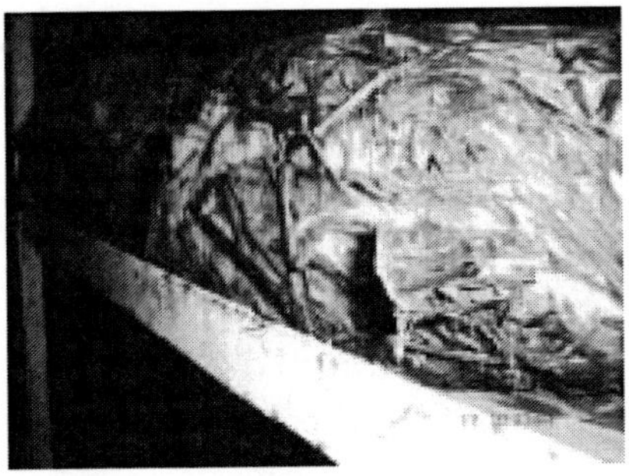

Fig. 21.

v) Fructification and Crop Management:

During spawn running period, water and light are not needed but a little ventilation is needed. By the end of 3-4 days, fluorescent light along with little more ventilation is provided in the rooms. The plastic sheets are removed on 4-5th day followed by little sprinkling of beds with water. The pinheads start appearing on 5th – 6th day of spawning (Fig. 22). After another 4 to 5 days, the first flush of mushroom will be ready for harvesting (Fig. 23).

The room conditions needed for better fructification are temperature 300C, relative humidity 80%, fluorescent light and intermittent fresh air. The quick growth rate of this mushroom demands ample supply of water and oxygen, which are antagonistic to each other in practice. Watering of the compost is not oftenly recommended as it lowers the temperature, suffocates the tiny primordia and reduces yield. Crop management to achieve the best possible combination of light, temperature, ventilation, relative humidity and compost moisture is in fact an art of judgment, experience and effort.

Fig. 22.

e) Circular Method

The steps involved in this method are :

- Preparation of bundles of one kg each from paddy straw.
- Soaking bundles in 2% $CaCO_3$ for 12 hours.
- Winding of water soaked bundles around wooden or cemented poles and mixing of mushroom spawn @ 1.5% on dry weight basis.

Fig. 23.

- Covering of spawned substrate with thin polythene sheet.
- Maintaining room temperature at 32-340C, RH 85%, with no light but little air circulation for 5-6 days.
- Removing of polythene sheet followed by spray of water on beds and lowering of room temperature to 28-320C, RH 80% and fluorescent light with enhancement in fresh air injection.
- Maintaining RH of 80% by fine misting on floor or walls without any direct spray on beds.
- Development of mushroom and harvesting by little lifting and twisting at egg stage.
- Packaging and sealing.

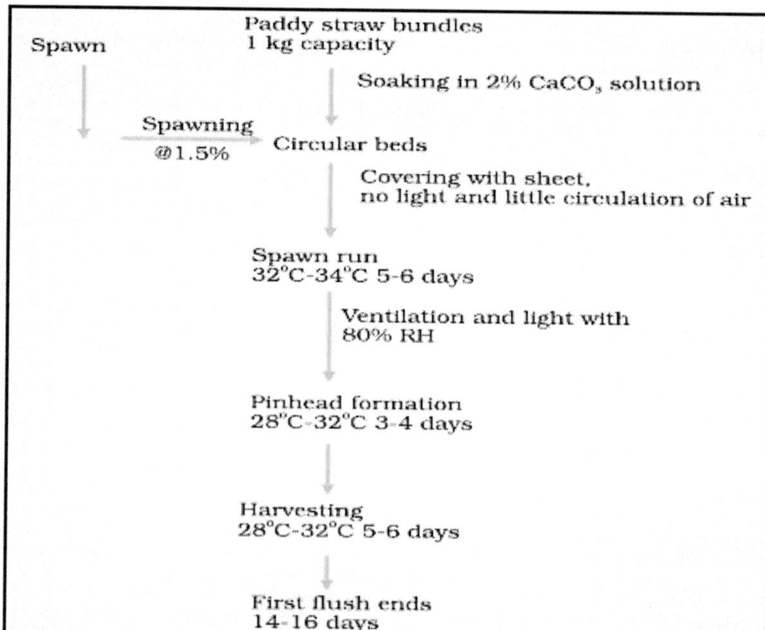

Fig. 24. Circular method of paddy straw mushroom cultivation

f) Indigenous Chinese Cultivation Practice

The method adopted at Green Poplar Village, Ping-Shan County, Hebei Province, China is mentioned below .

i) Compost preparation

- ❖ Overnight soaking of wheat straw (10-15cm long pieces) in 1% CaCO3 solution.
- ❖ Draining off of excess water by placing straw on ground.
- ❖ Piling of compost and covering with plastic sheet.
- ❖ Compost turning after 1 to 2 days interval, preferably on reaching the pile temperature at 500C.
- ❖ Filling of compost in 70cm x 35cm x 22cm size frame, first by putting a layer of compost followed by placing of spawn on four sides of this layer along with some wheat bran.

The second layer is placed on top of the first one and then spawn and wheat bran are added around the edges. The third and fourth layers are added like the first and second layers.

b) Arrangement of bed blocks

- ❖ Soil base is raised several centimeters surrounding the base of the frame.
- ❖ The blocks are arranged in two rows with a gap of 20-25 cm in between.
- ❖ Poplar branches are used to provide roofing on the blocks and are bowed in a shape to form the frame.
- ❖ Plastic sheet is spread over the frame, which in turn covered by straw.
- ❖ Temperature of around 33 to 350 C is maintained.

c) Harvesting of mushrooms

- ❖ Pinheads appear after 4-5 days of spawning.
- ❖ Total 9-10 days are taken for first harvest after spawning and the first flush lasts for 3 days accounting around 75% of the total mushroom yield.
- ❖ The bed blocks are watered with 0.5% CaCO3 and covered again.
- ❖ The second flush appears after few days and this flush accounts for rest 25% of the total mushroom yield.
- ❖ Four to 5 crops are harvested each year.

d) Spent compost

- ❖ The spent compost is dried and used for producing Pleurotus sajor-caju with BE of 80%.
- ❖ After P. sajor-caju cultivation, the spent compost can be used as a good soil conditioner.

Important steps for obtaining Healthy Mushroom Crop

- Compost moisture in the range of 60 to 65%.
- Immediate spawning on obtaining compost temperature at 350C followed by covering with plastic sheets, which should be maintained for next 4 days.
- No ventilation during first 3 days following spawning.
- Removal of plastic sheets after 4 to 6 days of spawning and sprinkling of water on bed surface followed by ventilating the cropping room.

Harvesting and Processing

Harvesting

The straw mushroom is harvested before the volva breaks or just after repture. These stages are called as the button and egg stages. This mushroom grows at high temperature and moisture, therefore, its growth is very fast.

So, for harvesting of straw mushroom at good condition it has to be harvested twice or thrice in a day (morning, noon & afternoon). This mushroom usually takes 9-10 days from spawning to first harvest of crop and the first flush normally lasts for 3 days, which constitutes about 70 to 90% of the expected mushroom yield. The intervening period of 3 to 5 days requires thorough watering and maintenance of optimum conditions inside the cropping rooms. The next flush will again lasts for 2-3 days and yields less mushroom than the first flush. The second flush adds only 10 to 30% of the total crop.

The mature fruiting bodies should be carefully separated from the beds/substrate by lifting and shaking slightly left or right and then twisting them off. The mushrooms should not be cut off by knives or scissors from the base of the stalk, because the stalk left behind on the bed/substrate will rot and be attacked by pests and contaminated by moulds, which in turn will destroy the mushroom bed.

Processing

Straw mushroom is more perishable than other edible mushrooms and can not be stored at 40C as it undergoes autolysis at this temperatureAhlawat et al. 2006). This mushroom can be stored at a temperature of 10 to 150C for 3 days and little more at 200C or under controlled atmosphere storage. The loss of moisture in 4 days stored mushroom could be as high as 40-50% in unpacked mushroom, while it can be reduced to 10% on packaging in perforated polythene begs. Straw mushroom can be processed by canning, pickling and drying.

However, practically the straw mushroom from China to Hong Kong is transported in wooden cases, in which the two compartments of the case are filled with ice, while the central compartment with mushroom.

On the other hand, this mushroom is transported by air from Taiwan to Thailand in bamboo baskets with central aeration tunnel and packed with dry ice wrapped in paper.

However, like button mushroom, more research work is needed in this mushroom also for studying the effect of blanching, post harvest storage, soaking and other chemical treatment before canning in order to increase the drained weight and improve the quality of the canned product.

Air Drying: Sun drying is very common in straw mushroom. The mushrooms are cut longitudinally before drying. Drying by hot air is better than sun drying because mushroom retains better flavour and colour.

Drying takes place in 24 hours at 300C. However, mushroom can also be dried with temperature beginning at 400C than increasing gradually until it reaches at 450C for eight hours.

Blanching of mushrooms for 3-4 minutes in hot water or 4-5 minutes in steam helps in retaining better colour of the dried product during storage.

Pretreatment of mushrooms with 0.1% KMS or combination of 0.05% KMS and 0.05% citric acid significantly improves the quality of the dehydrated product . The optimum drying temperature, time and critical moisture content for drying of the paddy straw mushroom has been recorded to be 600C, 7 hrs & 5%, respectively . Fresh mushrooms are reduced to about one-tenth of their original weight after dehydration. Dried mushrooms should be placed in air tight containers, to prevent moisture absorption. Dried mushrooms can be powdered and then used for making soup, ketchup or curry after reconstitution in water.

Freeze Drying: Freshly picked mushrooms are to be frozen at – 200C and then freeze dried. The finished produce on rehydration used to be better than air-dried product. On reconstitution it becomes almost indistinguishable in appearance from the fresh ones.

Conclusion

As we know cultivation of mushrooms and in particular the tropical mushrooms is the easiest way of agro-waste utilization in the shortest possible duration with an additional advantage of producing a quality food, possessing good proportion of essential amino acids, elements, fibre, ash and fatty acids. The additional advantage with paddy straw mushroom is, its shorter life cycle, fast growth, simple cultivation technique and high acceptability at consumers' level because of its unique texture and aroma. The bottle-neck of lower biological efficiency has almost been sorted out after bringing in of the cotton waste as the substrate; however, much more research work is needed to be done for developing suitable processing technology like of the button and other commercially grown mushrooms.

Chapter 7

Oyster Mushroom Cultivation

Introduction

Pleurotus species, commonly known as oyster mushrooms, are edible fungi cultivated worldwide especially in south east Asia, India, Europe and Africa. China produces 64 % of all edible mushrooms in the world and 85% of all oyster mushrooms all over the world (*Pleurotus* spp.) is also produced in China. Oyster mushrooms is the third largestcommercially produced mushroom in the world; however, Sánchezreported that *P. ostreatus* is the second largest next to *Agaricus bisporus* in the world market. Mushroom cultivation is the fifth largest agricultural sector in China with 24 billion USD value and 10% growth rate every year for the last 30 years. Oysters are naturally found on rotten wood material. The growing and consumption interest of oyster mushroom is increasing largely due to its taste, medicinal and nutritional properties. Large volumes of unused lignocellulosic byproducts are available in tropical and sub tropical areas. These by-products are left to rot in the field or are disposed off through burning. Utilizing these by-products for mushroom cultivation using locally available technologies may be one of the solutions to transforming these inedible wastes into accepted edible biomass of high market value.

Different species

❖ *Pleurotus sajor-caju*

❖ *Pleurotus florida*

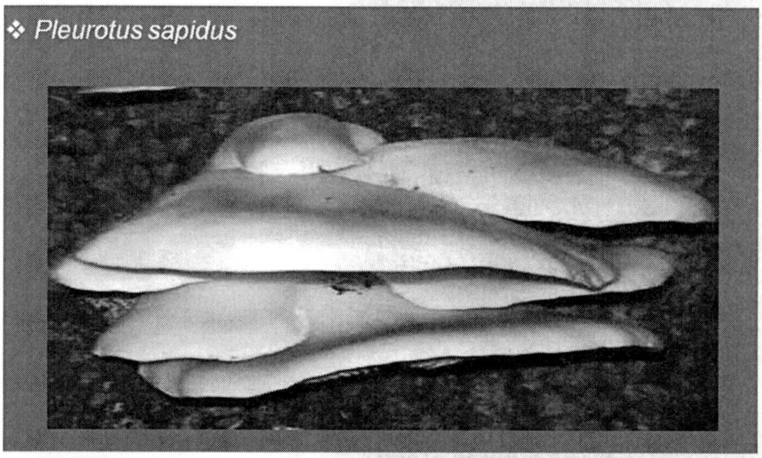

❖ *Pleurotus sapidus*

❖ *Pleurotus ostratus*

❖ *Pleurotus eous*

❖ *Pleurotus citrinipileatus*

❖ *Pleurotus flabellatus*

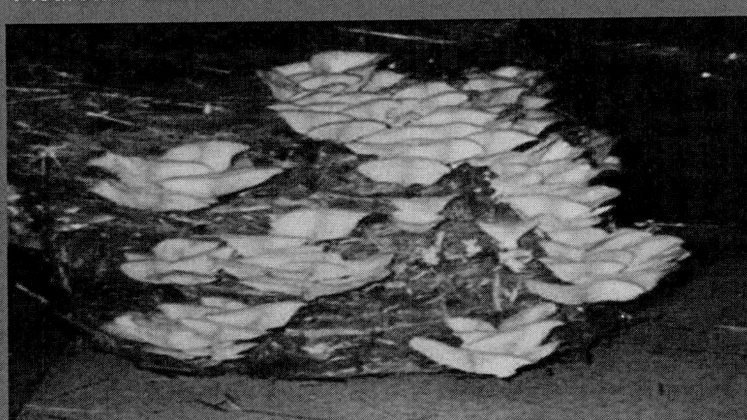

The spent substrates from mushroom cultivation can also potentially be used as an animal feed supplement, possibly providing additional animal feed resources . A shorter growth time is required to *P. ostreatus* in comparison to other edible mushrooms. The substrate used for their cultivation does not require sterilization, only pasteurization, which is less expensive.

High percentage of the substrates can be converted to fruiting bodies by oyster mushroom and hence it increases profitability. *P. ostreatus* demands few environmental controls, and their fruiting bodies are not often attacked by diseases and pests, and they can be cultivated in a simple and cheap way . All this makes *P. ostreatus* cultivation an excellent alternative for production of mushrooms when compared to other mushrooms. Therefore, it is better for unskilled farmers than other mushrooms. Mushroom cultivation provides an alternative employment and it contributes food security to rural disadvantaged groups especially women and old people in Tanzania hence improve their livelihood . The expansions of mushroom cropping decline the price of mushrooms and hence it safe guard food insecurity .

Debre Berhan is located in North Shewa Administrative Zone, Amhara Region and it is 130 Km North to the Capital City, Addis Ababa. Debre Berhan is the coldest and windy town in the country with an elevation of 2,850 meters above sea level.

According to World Bank , North Shewa is under high risk for drought and hence the zone needs an alternative agricultural practice that can't be easily affected by climate change.

Oyster mushroom cultivation is best in this regard since it doesn't require fertile soil and natural rain unlike other crops in Ethiopia. The aim of this research is to evaluate optimum conditions for oyster (*Pleurotus ostreatus*) cultivation using locally available substrates, materials and technologies.

Materials and Methods

Potato dextrose agar (PDA)

PDA was prepared from locally available fresh potato and industrially produced agar and glucose. About 200 g of potato was washed and sliced and placed in 1 liter of boiling water in a flask like brass locally available and boiled for 15 minutes. Potato broth was filtered using a piece of cloth. Glucose (20g) and 20g agar was added and the volume was adjusted to one liter by adding water. The flask was plugged with absorbent cotton. The agar mixture and petridishes were sterilized in pressure cooker for 30 minutes. The agar (25ml) was poured carefully and aseptically in to petridishes around flame and for comparison, the laminar air flow hood was also used. After cooling, the PDA was used for *P. ostreatus* inoculation and incubated at 25oC and at ambient temperature.

Spawn production

The grains, such as sorghum, wheat and dried malt wheat, were cleaned manually to remove inert matter, stubble and debris. The cleaned grains were soaked in tap water overnight. In addition, the grain was boiled soon after soaking for urgent spawn preparation. Thereafter, the soaked grains were drained and the excess water was removed and the following additives were added.

Wheat bran at the rate of 10%, and chalk (CaCO3) at the rate of 2% were added on dry weight basis of the grains. The additives were thoroughly and evenly mixed with the grains. The grain medium was filled in to bottles (made of brass) locally found. The bottles were covered using cotton wool plugged conduit. The bottles with spawn materials were autoclaved using pressure cooker and the supporting materials in bottles were allowed to cool for over six hours. The media in bottles were immediately inoculated with mycelial culture of *P. ostreatus* maintained on PDA. For successive production of spawn, the media in the bottles were inoculated with *P. ostreatus*. Then, it was incubated at 25oC for mycelial growth without any light for 10-15 days until the mycelium fully covered the grains. It was shaken every four days to distribute mycelium throughout the grain till the end of the growing day. Alternatively, the grains were put on bench at varying ambient temperature (12-19oC).

Substrate preparation

Wheat straw, barely straw, saw dust and sinar straw were collected from Debre Birhan surroundings and 90% of each substrate on dry basis were measured and soaked in tap water overnight to moisten and make the substrate soft for colonization to oyster mushroom hyphae. Alternatively 3% gypsum, chalk or CaCO3 was added to 10% wheat bran. The buffer and wheat bran were mixed separately and then they were added to the substrates.

In addition to different substrates, different supplements were also evaluated to observe their impact on oyster growth. Oil seed cake, wheat bran and cow dung (10% on dry basis) were added to 90% wheat straw and saw dust separately to ensure the even distribution of materials. Then the substrates and supplements were well mixed after the supplements and buffers were separately mixed. The relative humidity after mixing was measured as

$$\text{Relative Humidity} = \frac{\text{Wt of moistened substrate-wt of dry substrate}}{\text{Wt of moistened substrate}} \times 100\%$$

The growth of oyster on organic wastes

Many waste papers are released and burnt every month in Debre Berhan University. Hence an attempt to use waste paper as a medium for oyster cultivation was evaluated. Different amount of waste papers was mixed with saw dust. Saw dust (100%) was used as a control. In addition, in all 10% wheat bran was added after it was mixed with 3% gypsum separately. Finally the growth and the yield were measured using pileus diameter and biological efficiency. The *P. ostreatus* was inoculated and placed at ambient temperature. *Gabi* is, the Ethiopian traditional clothe and made of woven cotton, commonly dressed in the country. No clothe as frequently used as *gabi* by the majority of the people in the country and hence too much gabi wastes can be released every year in the country. Trials were made to cultivate the *P. ostreatus* on *gabi* wastes. Here 100%, 75%, 50%, 25%, and 0% saw dust were added to 0%, 25%, 50%, 75% and 100% gabi wastes respectively. In addition, 10% wheat bran and 3% gypsum were added. The percentage was calculated on dry basis. The substrates were pasteurized using oil drum. After cooling for an hour, the substrates was inoculated with *P. ostreatus* and placed at ambient temperature.

Reuse of the Substrates

The substrates that were used for cultivation of oyster for 2 flushes were used again for other cultivation. Here substrates were utilized in two ways. On the first hand, substrates without adding anything on used substrates were taken to grow mushrooms. On the other hand, 10% wheat bran and 2% gypsum was added to the used substrates. In addition, they were pasteurized in steam and sterilized using autoclave to evaluate the effect of contaminants that were presented in used substrates.

Spawning

The wooden hood and hands were disinfected with 70% alcohol. The mixer rod was sterilized by flame locally made. Then, the pasteurized substrates were inoculated with 4-6% *P. ostreatus* seed aseptically from mother spawn and they were incubated at 25oC and ambient temperature.

Effect of pore size

After substrates were filled to plastic bags, different size holes were made to evaluate effect of aeration, contamination and moisture loss. Pine holes, 16.18mm2 and 28.16mm2 holes were made using circular cutter with different diameter. Here pine holes were made after substrates were filled in the bags. However 16.18mm2 and 28.16mm2 holes were made before substrate were placed to the containers. The one with pin hole was arranged in two ways. First it was tied at top and second cotton plug was attached to plastic at the top to allow more aeration.

Effect of temperature

Since Debre Birhan is relatively cold especially from end of September to January, effect of ambient temperature on yield of oyster was evaluated since farmers in rural area are not able to buy incubators. Here, the bottle method was used due to its easiness to handle in the incubator. Equal amounts of substrates were put into two bottles. After inoculation, one was placed in the incubator at 25oC and the other was put on bench. The bench temperature was varying from 3oC to 19oC. Then every week the growth and development of oyster mycelium was physically observed and assessed.

Pinhead initiation

Light and temperature were used to initiate the formation of pinheads after the mycelium was fully grown on the substrate i.e. after the substrate was fully turned into white color due to mycelium mass. The growing mycelium that was placed in dark was brought to a fully illuminated region; on the other hand, the bottles and plastic bags that were put in the incubator at 25oC were placed at ambient temperature. The ambient temperature, when the experiment was conducting, was varying from 3oC to 19oC.

Fruiting

When the substrate was completely covered by the mycelium, the cotton plug was removed . Top of the plastic bag was partially removed. In addition, the part of the plastic that covered the primordia was also removed without harming the mycelium. The plastic bags that contain primordia was sprayed with water three times a day to increase the relative humidity .

Biological efficiency (yield)

Biological efficiency is a measureable tool used to know the growth potential of the oyster mushroom. It is calculated based on the following formula.

$$\text{Biological efficency} = \frac{\text{Weight of fresh mushroom}}{\text{Dry weight of substrate}} \text{X } 100\%$$

Results and Discussions

Oyster culture on PDA

PDA is the simplest and the most popular medium for growing mycelia of most cultivated mushrooms. *P. ostreatus* was successfully grown on PDA. The oyster completely covered the petridishes in 9 days and its color and appearance looks like pure cotton (fig 1). The mycelium should be white and grow out from the tissue. If yellow, blue, green or grey mycelia form on other places on the surface, then these are fungal contaminants. A creamy, shiny growth often indicates bacterial contamination. *P. ostreatus* is a slow grower when it is compared to molds and other fungi.

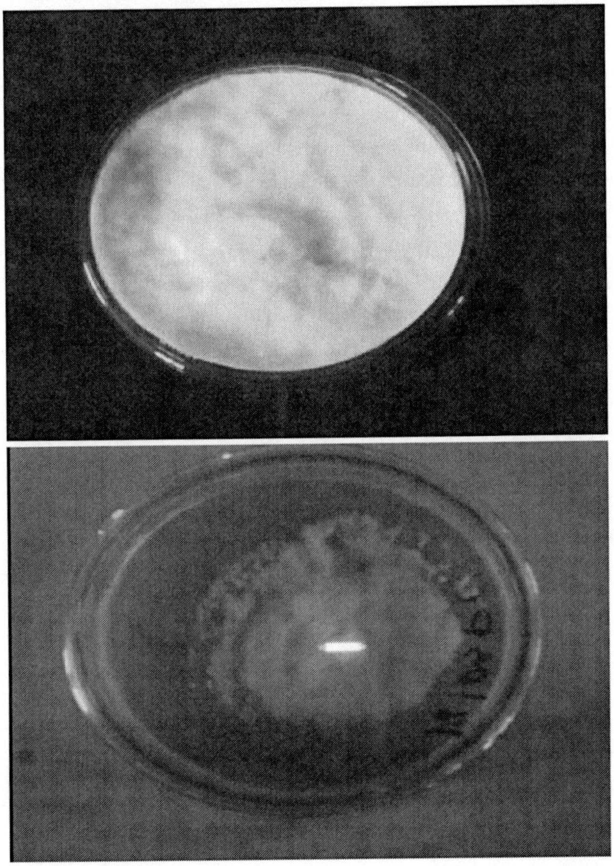

Fig. 1: ***P. ostreatus*** **culture on PDA 5 days old (right) and 11 days old (left).**

Oyster spawn

Once a pure culture of mushroom is obtained, the spawn could be made from it. The spawn is the mycelium of the mushroom on a solid material. Functionally, it is the starter inoculums of mushroom cultivation. Solid materials such as saw dust, whole grain of cereals, shredded maize and corn cobs etc are used for spawn preparation .

P. ostreatus was grown on sorghum wheat, and dried malt wheat based spawn without showing significant preference on them. Sorghum and wheat are cheap in Debre Birhan market hence it is economical to use them to produce spawns. The choice of grain is made after considering several factors such as its prevailing price, easy supply and grain size. In some areas in Africa, sorghum is better available and cheaper than maize .

Besides, the space between wheat or sorghums is quite enough for air circulation. However if teff was used, there would be no good aeration and teff seeds also get clamped together during sterilization . That is why oyster didn't grow very well in spawn that contain too much wheat bran especially at the bottom of the flask (data not shown).

The whole inoculated cereals put on incubator at 25oC were covered by mycelium of *P. ostreatus* within 14 days.

However, those placed at bench took 22 days to colonize the cereal grains. Therefore, it is possible to prepare spawn without incubator, by simply placing on bench even if it takes longer period relatively.

Sorghum needs longer time to imbibe the same amount of water than wheat . That is why sorghum was boiled but the sorghum should not be burst because the burst sorghum will be clamped and then it reduces the air circulation due to starch exposure during bursting.

Oeicompared the advantage and disadvantage of grains over sawdust or wooden stick. The main advantage of grains is that it is very nutritious for fungi and form kernels easily. The kernels can easily be distributed to the substrate. The main disadvantage is that it provides optimal substrate for other organisms too. The chances of contaminations are therefore much higher compared to sawdust or wooden stick spawns.

The moisture content of the grain should be around 50% . However, Dawitstated that water content of 40-60% is optimal for spawn making. In our case, the water content was 52%. If the water content is higher, mycelia growth may be faster but the danger of wet spot bacteria is also greater. If it is drier than 35%, mycelia growth will be slow . To get good quality of spawn, what is most important is the quality of the inoculums. The inoculums should be fresh and pure. If the inoculums are preserved in refrigerator, they should be activated before they are used for spawn production.

Substrates

Mushroom cultivation is different from conventional agriculture in one major point. The soil in the ground is the substrate for production of crops but the mushrooms grow on lignocellulosic agro industrial wastes.

Oyster can grow on wide agricultural wastes what matter is which agricultural residues are found around us.

Wheat straw, barely straw, sinar straw and saw dust were used as substrates for *P. ostreatus*. There was no significant biological efficiency difference on the substrates. It was successfully grown on them. Dawitpointed out that *Pleurotus* can grow on wheat and barley straws.

The moisture content of the substrates was varying from 69.8% to 74.5%. Significantly the highest run rate was recorded at 70% moisture level. pH of the straw before adding chalk and gypsum was 5.8 .This might be because of the acids produced by microbes in substrates during soaking of substrates in water since it was soaked overnight.

However, the pH rose to 6.9 when chalk was added. Here, washing is essential to neutralize the pH of the substrate. In addition, the chalk neutralizes the lowered pH.

Furthermore, chalk and gypsums act as buffer. Besides, the Ca2t obtained from gypsum (CaSO4) and chalk (CaCO3) neutralize oxalic acid produced during mycelial growth. Highest yield is observed at pH 7 even though there is no significant yield difference at the pH 6-8. However, the biological efficiency, biological yield and economic yield were increased with the increase of pH levels up to 5.04 and then decreased .

On contrary, Nwokoye *et al.*showed that *P. ostreatus* was able to grow optimally at pH of 9. Oyster productivity difference when different buffer such as gypsum, chalk and pure CaCO3 were optimized; they showed no difference on oyster yield. However, there was pH difference after they were added.

Supplements are additives which increase the yields by providing specific nutrients for the growth of the mycelium . However, supplements increase the risk of contamination at least by 25%since supplements also provide good nutrients to other microorganisms. Dawitalso stated that supplements change physical conditions of substrates more suitable for cultivation of mushrooms. Addition of supplements (rice husk) to waste paper significantly increase spawn running, pin head formation, fruit body formation and mushroom yield .

Similarly, cultivation of oyster mushroom on wheat straw and bagasse amended with distillery effluent yield better result than substrates without distilleryeffluent. The effects of wheat bran, oil seed cake (fig 2) and cow dung on growth were evaluated and it was found that oil seed cake gave maximum yield (table 1 and fig 2). Likewise, Ruiz-Rodriguez *et al.*found that oyster yield increased on substrates with supplementation of oil mill waste without affecting cultivation parameters. The yield reduction in cow dung might be due to contamination since pasteurized substrates containing caw dung was partly contaminated by green mold.. Similarly, Baysal *et al.*proposed that the yield reduction in oyster production during supplimenattion of peat and chicken manure might be due to high nitrogen content of the substrate.

Fig. 2: Supplements: wheat bran (left) and oilseed cake of niger seed (right)

The substrates used for cultivation of oyster were used again for other cultivation. The pasteurized substrates were found to be contaminated. However, *P. ostreatus* was grown in noncontaminated parts. As it is indicated in table 1, reused substrates that contained additional supplements exhibited better growth on both pasteurized and sterilized substrates. The yield decrease in reused substrate is because *P. ostreatus* is primary decomposer.

The reused substrates were very soft and made into small pieces easily. Therefore, the aeration in reused substrates will be relatively lower than original substrates. Furthermore, biological efficiency of oyster was evaluated using waste paper and gabi wastes and it was found that waste paper and gabi were better than wheat straw (fig 3).

Fig. 3: Oyster in gabi waste alone (left) and saw dust with *gabi* waste (right).

In order to find why gabi and waste paper gave maximum yield, moisture holding capacity of the substrates were evaluated and the waste paper and gabi had high water holding capacity. Hence one reason for high yield of oyster in gabi and waste paper might be its high water holding capacity. Oyster cultivation on shredded office paper and cardboard yielded more edible sporophore biomass than other lignocellulosic residues .

Effect of pore size

Pin holes, 16.18 mm2 and 28.16mm2 holes were made on plastic bags to evaluate the effect of aeration, contamination and moisture loss. Substrate in plastic with pin holes and 16.18 mm2 holes were well colonized by *P. ostreatus*. However, the yield was lower in bags with 16.18 mm2 (table 1). On plastic bag with 28.16mm2 areas around the holes were not colonized and it was contaminated by molds. Inability to colonize might be due to great loss of moisture since the hole is large enough.

Effect of temperature

The substrates in the bottle placed in the incubator at 25°C were colonized by oyster within 15 days where as the one put at beach with temperature varying from 3 to 19°C was colonized within 21 days. At the temperature below 14°C, the growth of the oyster was found to be so slow that they couldn't give primordial within 28 days. In stead 46 days were required for pinhead formation at temperature below 14oC and the yield was also very small (table 1). To increase the temperature, the room was heated using wood charcoal. For most mushrooms, mycelial invasion is favored at higher temperature and fruit body formation at relatively lower . On the other hand, Alam *et al.*investigated that the maximum growth is recorded at 25 °C and the lowest at 15 °C and no significance difference is observed between 25 °C and 30 °C. Therefore, the temperature ranging between 25- 30 °C is best for mycelial growth of *P. ostreatus*. However, edible mushrooms cultivated between 15-20 °C present better quality and durability than those at 25°C. In addition, *P. ostreatus* grew faster at 30 °C . In general, oyster yield decreases when the temperature decreases in different climatic zones of Pakistan .

Pin head initiation

Light and temperature changes were used to facilitate pinhead formation. Humidity and CO_2 concentration adjustment is also essential. All oyster mushrooms need light for the proper development of cap. Usually there is enough light for oysterif it is possible to read a news paper in the growing room. Pin heads were developed in 26-31 days. *P. ostreatus* completed spawn running in 17-20 days on different substrates and time for pinheads formation was noted as 23-27 days . In addition, temperature plays very important role for pin head formation. That is why pinhead evolvements were failed at a temperature lower than 11°C.

Furthermore, humidity should be very high during primordial formation.

To achieve it, the mycelium was sprayed with water 3 times a day. Oeistated that humidity at time of induction must be ≥ 90%. In some countries like Philippines and Nepal growers take off all or almost all the plastic from the bags. This leads to fast drying out of the substrate and lower the yields and hence it should be discouraged . Another disadvantage of removing plastic is that more but smaller fruit bodies are formed, which increases the picking cost considerably.

Harvesting

At the first trial, only the first flush was harvested and 2nd and other flush were failed. These motivated us to see different environmental conditions and when the mycelium were covered by plastic after spraying with water, pinheads were developed and then 2nd, 3rd and 4th flushes were also successfully cultivated. Hence, what is most important for 2nd, 3rd and 4th flushes cultivation is humidity. Furthermore, temperature also affected our 2nd, 3rd, and 4th flushes.

Unlike many other mushrooms by which the yield continuously decrease in consequent flushes, the pattern of *P. ostreatus* is different. More yields were found in second flushes (table 1). The highest crop of the mushroom is obtained during the second and third flush. However, the maximum yield was obtained in first flush than the second and third flush. In general, one crop of mushroom with four flushes is most economical. It is unwise to keep spawn running of substrate for longer than 90 days as the yield significantly decrease.

Good control of the humidity during cropping is very important for all types of mushroom. The moisture content of growing mushroom media is a very important factor; hence, the proper moisture content value encourages the growth, while higher or lower ones had a negative effect on growth. It is good keep the humidity high (80 - 90%) by spraying water several times per day. However, no water should be sprayed directly onto mushrooms that are ready for picking; their shelf life will decrease drastically if they become too wet.

Table 1: Oyster yield with varying substrates, supplements, pore sizes, temperature, reused substrate and flushes.

Experimentation	BE%	Experimentation		BE%
Substrates		Effect of temperature		
Wheat straw	68.16	Temperature	Primordia days	BE%
Sinar straw	66.02	3 to 19°C (ambient)	46 days	65.44
Barely	65.52	25°C	28 days	71.59
Saw dust	66.31	Effect of reused substrate with different heat treatment		
Supplements		Reused substrates	Heat treatment	BE
Wheat bran	64.21	Used substrate	Pasteurization	21.54*
Oil seed cake (Niger seed)	71.56	Used substrate and supplements	Pasteurization	28.18*
Cow dung	12.32*	Used substrate	Sterilization	39.39
Pore size		Used substrate and supplements	Sterilization	54.02
Pin holes	62.31			
16.18mm^2	60.02			
28.16mm^2	28.79*			
Flushes				
First flushes	18.24			
Second flushes	24.67			
Third flushes	16.92			
Fourth flushes	10.22			

*Was found to be contaminated

Table 2: Effect of buffer on wheat straw substrate.

Buffer	Initial pH	pH after addition of buffer	Biological efficiency
Gypsum	5.8	6.87	59.23
Chalk	5.8	6.93	60.19
Pure Na_2CO_3	5.8	7.02	58.77

Conclusions

Relative humidity, aeration, temperature, and contaminations are most important factors during oyster cultivation in Debre Berhan using locally available substrates, materials and technologies. Even though oyster spawning grow best at 25oC, it is possible to cultivate it at lower temperature.

Drying the substrates greatly reduce the yield hence spraying it with water is mandatory. To activate 2nd, 3rd, and 4th flush, covering the substrates with plastic is very important to make the substrate moist. Gabi and paper wastes can be used as a substrate separately or in mixture with saw dust. Generally, oyster can be cultivated in Debre Berhan with locally available materials without using sophisticated lab equipment for spawn production.

Chapter 8

Growing Oyster Mushrooms on Toilet Paper Rolls

Materials

- ❖ Oyster spawn (either grain spawn or sawdust spawn)
- ❖ Large filter patch bags
- ❖ Toilet paper rolls (Generic white rolls of toilet paper work best. Avoid kinds that have color printing, added scent, or lotions.)

1 **Dip the rolls.** Bring a pot of water to a boil. Either dip your rolls of toilet paper into the boiling water using tongs, or pour the boiling water over the rolls of toilet paper in a sink. This is a "soft" (i.e., not strictly complete) pasteurization of the roll. An oven rack placed over the sink makes a good makeshift rack to hold the rolls.

2 **Inoculate the rolls.** Allow the toilet paper roll to cool on a small plate. Crumble up your Oyster spawn and fill the interior of your toilet paper roll with it. We generally also dust the exterior of the roll with some spawn. Enclose the roll and plate in a filter patch spawn bag. Use a rubber band at the top of the bag to close it.

1

2

3 **Air and mist.** Open the bag several times a day to air it out and mist it with a spray bottle. Keep the sides of the plastic bag moist by making them the primary target of the spray bottle. Try to avoid water pooling in the bottom of the bag. If you need to, you can go without misting for a day or two without negative consequences.

4 **Harvest.** Your roll should be fully colonized in 2 weeks or less if you're using Oyster mushrooms. Your fruits should come a week or two after full colonization. To harvest, simply grasp the mushroom at the base of the fruit body and gently twist it off with your fingers. You can sometimes get two or three flushes from this method as long as the roll doesn't dry out. Subsequent flushes should start around 1 week after the previous harvest.

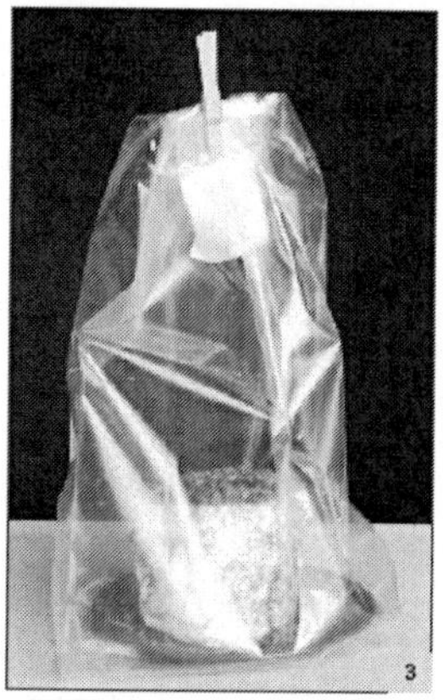

3

4

Chapter 9

Cultivation of Oyster Mushroom on different agricultural waste substrate and its nutrient analysis

Mushroom with their great variety of species, constitute a cost-effective means of both supplementing the nutrition to human kinds. 4-5 species of mushroom are of industrial significance throughout the world . In India, only 3 species, namely, *Agaricus bisporus*, *Pleurotus sajorcaju* and *Volveriell* are preferred for commercial cultivation. Of the three cultivated species, the white button mushrooms have the highest consumer preference and account for about 90 per cent of total mushroom production. The Oyster mushroom grows during winter months only therefore, it needs proper preservation techniques to promote their consumption among common people and excess mushroom is processed into food products acceptable to consumers. Mushrooms contain 90 per cent moisture. Oyster mushroom (*Pleurotus* sp.) belonging to Class Basidiomycetes and Family Agaricaceae is popularly known as 'Dhingri' in India and grows naturally in the temperate and tropical forests on dead and decaying wooden logs or sometimes on dying trunks of deciduous or coniferous woods. It may also grow on decaying organic matter. Produces protein rich food Oyster mushrooms are the third largest cultivated mushroom. The economic importance of the mushroom lies primarily in its use as food for human consumption. It is rich in Vitamin C and B complex and the protein content varies between 1.6 to 2.5 percent & mineral salts required for the human body. The niacin content is about ten times higher than any other vegetables.

The folic acid present in oyster mushrooms helps to cure anemia. It is suitable for people with hyper-tension, obesity and diabetes due to its low sodium: potassium ratio, starch, fat and calorific value. Alkaline ash and high fiber content makes them suitable for consumption for those having hyperacidity and constipation & cholesterol inhibitors mushrooms. Mushrooms are rare vegan sources of vitamin D and conjugated linoleic acid. Mushrooms have antioxidant property due to presence of compounds like Ergothioneine. Oyster mushroom can grow at moderate temperature ranging from 20 to 300 C and humidity 55-70% for a period of 6 to 8 months in a year. It can also be cultivated in summer months by providing the extra humidity required for its growth in hilly areas above 900m. (m.s.l.), the best growing season is during March/April to September/ October and in the lower regions from September/October to March/April. In this study we are going to observe growth of same species on different raw material, Nutritional qualities of same species on different raw material and physico-chemical characteristics of oyster mushroom.

Classification of Oyster Mushroom-

Scientific classification-

Kingdom – Fungi

Phylum – Basidiomycota

Class - Agaricomycetes

Order - Agaricales

Family - Pleurotaceae

Genus – *Pleurotus*

Species - *ostreatus*

Materials and Methods

Materials-

1. Spawn (Source:- Agriculture College, Pune)
2. Paddy Straw, Wheat Straw as raw materials.
3. Water bath
4. Hot Air Oven
5. Plastic Bags

Chemicals-

1. 1N NaOH
2. Chloroform : Methanol(2:1)

Methods-

Oyster Mushroom Cultivation

Substrate preparation

Oyster mushroom was grown on various substrates viz., paddy straw, Wheat straw, vegetable plant residues etc.

Since paddy straw is easily available and cheap, it is widely used. Paddy straw used was fresh and well dried.

Soaking

Wheat and paddy straw were chopped into 3-5 cm pieces and soaked in fresh water for 8-16 hours. Excess water from straw was drained off by spreading it on filter paper.

Heat Treatment

Heat treatment of substrate results in minimizing contamination problem and gives higher and almost constant yields. It can be done e by pasteurization.

Pasteurization

Water was boiled in a wide mouth container such as tub or drum. The wet substrate was filled in gunny bags. The filled bag was dipped in hot water of 80-85C for about 10-15 minutes. To avoid floating, it was pressed with some heavy material or with the help of a wooden piece. After pasteurization, excess hot water was drained off from container so that it can be reused for other sets & hot water temperature was maintained at 80-85C for all sets to achieve pasteurization.

Spawning

When the pasteurized substrate had cooled down to room temperature, it was ready for filling and spawning. At this stage, substrate moisture content was about 70%. Polythene bags (35 x 50 cm, 150 gauges) or polypropylene bags (35 x 50 cm, 80 gauges) were used for its cultivation. One 500 ml bottle spawn (200-250 g) can be used for 10-12 kg wet straw (3 bags). Spawning can be done in layer spawning or through spawning.

In case of layer spawning, substrate was filled in bag, pressed to a depth of 8-10 cm and broadcasted with a handful of spawn above it. Similarly, 2nd and 3rd layers of substrate were put and simultaneously after spawning, the bags were closed. After spawning, pasteurized straw was mixed with 2% spawn and filled in bags. After that it was gently pressed, and the bags were sealed for spawn running (development).Spawned bags were stacked on racks in neat and clean place, in closed position. Temperature at 25-35 C and humidity at 70-85% was maintained by spraying water twice a day on walls and floor. It took 15- 20days when bags were fully covered with white and pink mycelium respectively.

Cropping and harvest

After 20-22 days, when bags were fully impregnated with white mycelium, they were transferred into cropping room and the polythene covers were removed. The open blocks were kept on racks about 20 cm apart with gap of 50-60 cm between two shelves. Mushrooms were grown in a temperature range of 20-33 C. Relative humidity was maintained by spraying water twice a day on the walls and floor of the room. A light spray of water was given on blocks as soon as the small pin heads appeared. Once pinheads were 2-3 cm big a little heavier watering was done on blocks and watering of blocks was stopped to allow them to grow. Mushrooms were plucked before they shed spores to maintain quality. After 1st flush of harvest, 0.5 to 1 cm outer layer of the block was scrapped. This helped to initiate 2nd flush which appeared after 10 days.

After harvesting they were packed in perforated (5-6 small holes) polythene bags to keep them fresh. It loses freshness after about 6 hours, which can be enhanced by keeping them in refrigerator. Oyster mushroom can be shed dried for 2 days and dried product marketed in polythene bags. Dried mushrooms were soaked in water for 10 minutes before use.

After the growth of *oyster* mushrooms following biochemical test were carried out –

Determination of Moisture

The moisture content is determined by measuring the of a material before & after the water removed by evaporation

Moisture content was determined by following –

%Moisture =initial –dried/initial *100

Here M initial & M dried are the mass of sample before & after drying respectively To obtain an accurate measurement of the moisture content of material evaporation method necessary to remove all water molecules.

Determination of total protein-

Protein content was measured by using Biuret method.

Determination of total lipid-

Total lipid was determined by slight modified method of Folch et al. Five gram of each sample was suspended in 50ml of chloroform: methanol (2:1) mixture then mixed thoroughly and let stand for 3 days. The solution was filtrated and centrifuged at 1000g by a centrifuge machine. The upper layer of methanol was removed by Pasteur pipette and chloroform was evaporated by heating. The remaining was the crude lipid.

Determination of total ash-

One gram of the sample was weighed accurately into a crucible. The crucible was placed and heated first over a low flame till all the material was completely charred, followed by heating in an oven for about 6 hours at 6000C. It was then cooled and weighed.

Then total ash was calculated as following equation :

Ash content (g/100g) = weight of ash *100/ weight of sample

Determination of crude fiber-

Crude fiber determined using Soxhelt extraction method.

Determination of total carbohydrate-

The content of the available carbohydrate was determined by the following equation :

Carbohydrate (g/100g sample) = [100 – (Moisture + Fat + Protein + Ash + Crude Fiber)]

Determine Metabolizable energy-

The content of availed or carbohydrates can supply energy.

ME (Kcal/100g) = [(3.5*CP)+(8.5*CF)+(8.5*NFE)]

Where

ME = metabolic energy CP =crude protein

CF = crude fate (lipid) NFE = % nitrogen free extract(carbohydrate)

Determination of calcium and magnesium content:

1) Estimation of total hardness(Ca & Mg):

- Take 25 ml of sample in a conical flask & add 1ml of buffer solution.
- Add a pinch of Erichrome black T indicator in it, The solution turns red
- Titrate the whole content against 0.01M EDTA solution.

- Take 3 readings; the end point is wine red to blue.
- Note the mean burette reading, and calculate the total hardness using following formula –

Total Hardness (mg/ml) = MBR*M*1000/Volume of sample (ml)

2) Estimation of Conc. of Calcium :

- Take 25ml of sample in a conical flask and add 2ml of 1N NaOH to it.
- Add a pinch of Muroxide indicator to it which turns solution into pink.
- Titrate the whole content against 0.01 M EDTA solution.
- Take 3 readings; the end point is pink solution turns to purple.
- Note the mean burette reading, and calculate amount of Calcium in the sample using following formula

Conc. of Calcium (mg/ml) = MBR*M*1000/Volume of sample

Conc. of Magnesium (mg/ml) = Conc. of total hardness - Conc. of Calcium

I. Grey Mushroom

a) Substrate: Wheat Straw

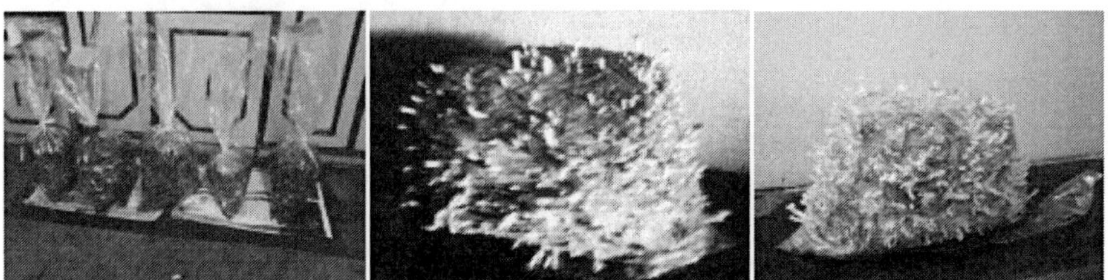

DAY 1: Spawning & Bagging DAY 12: Initial budding DAY 13: Initial fruiting

DAY 16: Initial growth DAY 17 : Grown stage DAY 18: Full grown grey Mushroom

b) Substrate: Paddy Straw

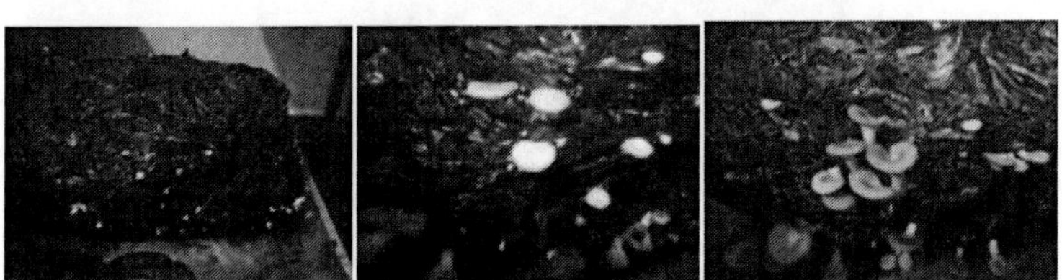

DAY 13: Budding DAY 15: Initial fruiting DAY 17: Initial growth

DAY 18: Full grown grey Mushrooms

DAY 20: Visible gills of mushroom

PINK MUSHROOM-

a) Substrate: Wheat Straw

DAY 12: Initial Budding

DAY 14: Initial Fruiting

DAY 15: Fruiting

DAY 17: Initial Growth

DAY 18: Grown mushrooms

DAY 20: Full grown flower Shaped mushroom

b) Substrate: Paddy Straw

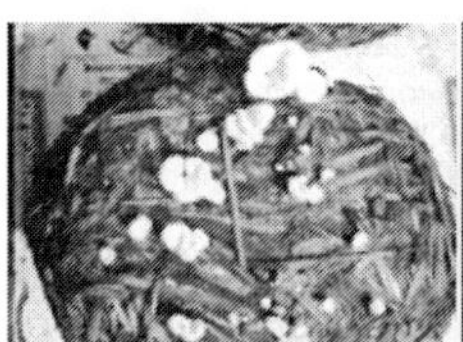

DAY 17: Budding

DAY 18: Initial Fruiting

DAY 19: Fruiting

DAY 20: Full grown mushrooms

DAY 21: Visible gills seen in mushroom

- **Determination of protein content -**

Test Tube No	Vol. of BSA (ml)	Vol. of Saline (ml)	Vol. of $CuSO_4$ (ml)	Stand For 15Min. At R.T.	Vol. of FC reagent (ml)	Stand For 30 Min. At R.T.	Conc. Of BSA (μg)	O.D at 660 nm (gray wheat straw)	O.D at 660 nm (gray paddy straw)	O.D at 660 nm (pink wheat straw)	O.D at 660 nm (pink paddy straw)
Blank	0.0	1.0	5		0.5		00	0.0	0.0	0.0	0.0
1	0.2	0.8	5		0.5		20	0.20	0.20	0.20	0.20
2	0.4	0.6	5		0.5		40	0.33	0.33	0.33	0.33
3	0.6	0.4	5		0.5		60	0.47	0.47	0.47	0.47
4	0.8	0.2	5		0.5		80	0.60	0.60	0.60	0.60
5	1.0	0.0	5		0.5		100	0.74	0.74	0.74	0.74
6	0.5 (Sample)	0.5	5		0.5		-	0.81	0.13	0.50	0.17

- **Determination of Sugar Content-**

Test Tube No.	Std. glucose (ml)	Disstelled Water	Phenol	Conc. H_2SO_4 (ml)	Stand For 30 Min. At R.T	Conc. Of Glucose	O.D at 660 nm (gray wheat straw)	O.D at 660 nm (gray paddy straw)	O.D at 660 nm (pink wheat straw)	O.D at 660 nm (pink paddy straw)
Blank	0.0	1	1	5		00	0.0	0.0	0.0	0.0
1	0.2	0.8	1	5		20	1.06	1.06	1.06	1.06
2	0.4	0.6	1	5		40	1.19	1.19	1.19	1.19
3	0.6	0.4	1	5		60	1.42	1.42	1.42	1.42
4	0.8	0.2	1	5		80	1.50	1.50	1.50	1.50
5	1.0	0.0	1	5		100	1.63	1.63	1.63	1.63
6	0.5 (Sample)	0.5	1	5		-	0.18	1.52	1.30	0.99

- **Determination of Moisture Content-**

Mushroom	W_1 (Wt. of mushroom before drying)	W_2 (Wt. of mushroom after drying)	Moisture Content $(W_1 - W_2)/W_1 * 100$
Gray Wheat Straw	6	0.75	0.875 = 87.5%
Gray Paddy Straw	20	4.5	15.5 = 77.5%
Pink Wheat Straw	8.660	0.690	0.92 = 92%
Pink Paddy Straw	18.00	2.5	0.8611 = 86.11%

Determination of Crude Fiber -

Mushroom	W_1= Initial Wt. of beaker (in gm)	W_2 = Final Wt. of beaker (in gm)	Crude Fiber $W_2 - W_1$ (in gm)
Gray Wheat Straw	111.0	111.81	0.81
Gray Paddy Straw	100.750	101.560	0.81
Pink Wheat Straw	102.0	103.260	1.260
Pink Paddy Straw	98.950	99.040	0.09

- **Determination of Lipid Content-**

Mushroom	W_1= Initial Wt. of beaker (in gm)	W_2 = Final Wt. of beaker (in gm)	Lipid content $W_2 - W_1$ (in gm)
Gray Wheat Straw	96.870	97.580	0.71
Gray Paddy Straw	141.20	142.47	1.27
Pink Wheat Straw	150.260	151.260	1
Pink Paddy Straw	157.250	157.620	0.37

Determination of Ash Content-

Mushroom	W1= Initial Wt. (in gm)	W2 = Wt. of Ash (in gm)	Ash content W2 / W1*100 (in gm)
Gray Wheat Straw	5	4.090	81.8
Gray Paddy Straw	10	5.730	57.3
Pink Wheat Straw	10	0.740	7.4
Pink Paddy Straw	10	5.73	57.3

RESULTS AND DISCUSSION

1. GREY MUSHROOM

A) Substrate- Wheat Straw

- **Determination of Calcium and Magnesium Content-**

Part I-Estimation of Total Hardness

Readings	Burette Readings			MBR
	I	II	III	
Final	9	9	9	9
Initial	0	0	0	
Difference	9	9	9	

Calculations:
Total Hardness (mg/l)
= MBR*M*1000/ Vol. of sample (ml)
= 9*0.01*1000/25 = 3.6 mg/l

Part II- Estimation of Conc. of Calcium

Reading	Burette Reading			MBR
	I	II	III	
Final	1	1	1	1
Initial	0	0	0	
Difference	1	1	1	

Calculations:
Conc. of Calcium (mg/l)
= M.B.R *M* 1000/Vol. of sample (ml)
=1*0.01*1000/25 = 0.4 mg/ml

Conc. of Magnesium (mg/l) = Conc. of total hardness- Conc. of Calcium
= 3.6 - 0.4 = 3.2 mg/ml

- **Determination of metabolise energy-**

ME (Kcal/100g) = [(3.5*C.P) + (8.5*CF) + (8.5*NFE)]
= [(3.5*10.3) + (8.5*0.71) + (8.5*1.6)] = 55.685

B) Substrate - Paddy Straw

- **Determination of Calcium and Magnesium Content-**

Part I- Estimation of Total Hardness

Reading No.	Burette Reading			MBR
	I	II	III	
Final	11.8	11.5	11.5	11.5
Initial	0	0	0	
Difference	11.8	11.5	11.5	

Calculations:
Total Hardness (mg/l)
= MBR*M*1000/ Vol. of sample (ml)
= 11.5*0.01*1000/25
= 4.6 mg/l

Part II- Estimation of Conc. of Calcium

Reading	Burette Reading			MBR
	I	II	III	
Final	3	3	3	3
Initial	0	0	0	
Difference	3	3	3	

Calculations:
Conc. of Calcium (mg/l)
= M.B.R *M* 1000/Vol. of sample (ml)
=3*0.01*1000/25
= 1.2 mg/ml
Conc. of Magnesium (mg/l) = Conc. of total hardness- Conc. of Calcium
= 4.6 - 1.2 = 3.4 mg/ml

- **Determination of metabolise energy -**

ME (Kcal/100g) = [(3.5*C.P) + (8.5*CF) + (8.5*NFE)]
= [(3.5*1.6) + (8.5*1.27) + (8.5*13.2)] = 128.595

2. PINK MUSHROOM

A) Substrate – Wheat Straw

- **Determination of Calcium and Magnesium Content-**

Part I-Estimation of Total Hardness

Reading No.	Burette reading			MBR
	I	II	III	
Final	8.5	8.5	8	8.33
Initial	0	0	0	
Difference	8.5	8.5	8	

Calculations:
Total Hardness (mg/l)
= MBR*M*1000/ Vol. of sample (ml)
= 8.33*0.01*1000/25
= 3.32 mg/l

Part II- Estimation of Conc. of Calcium

Reading	Burette Reading			MBR
	I	II	III	
Final	3	2.7	3	2.8
Initial	0	0	0	
Difference	3	2.7	3	

Calculations:
Conc. of Calcium (mg/l)
= M.B.R *M* 1000/Vol. of sample (ml)
= 2.8*0.01*1000/25
= 1.12 mg/ml
Conc. of Magnesium (mg/l) = Conc. of total hardness- Conc. Of Calcium
= 3.32 - 1.12 = 2.2 mg/ml

- **Determination of metabolise energy-**

ME (Kcal/100g) = [(3.5*C.P) + (8.5*CF) + (8.5*NFE)]
= [(3.5*2.1) + (8.5*0.37) + (8.5*8.4)] = 81.895

B) Substrate – Paddy Straw

- **Determination of Calcium and Magnesium Content-**

Part I-Estimation of Total Hardness

Reading	Burette reading			MBR
	I	II	III	
Final	10	10	10	10
Initial	0	0	0	
Difference	10	10	10	

Calculations:
Total Hardness (mg/l)
= MBR*M*1000/ Vol. of sample (ml)
= 10*0.01*1000/25
= 4 mg/ml

Part II- Estimation of Conc. Of Calcium

Reading	Burette Reading			MBR
	I	II	III	
Final	1	1	1	1
Initial	0	0	0	
Difference	1	1	1	

Calculations:
Conc. of Calcium (mg/l)
= M.B.R *M* 1000/Vol. of sample (ml)
=1*0.01*1000/25
=0.4 mg/ml
Conc. of Magnesium (mg/l) = Conc. of total hardness- Conc. of Calcium
= 4 - 0.4 = 3.6 mg/ml

- **Determination of metabolise energy-**

ME (Kcal/100g) = [(3.5*C.P) + (8.5*CF) + (8.5*NFE)]
= [(3.5*2.1) + (8.5*0.37) + (8.5*8.4)]
= 81.895

Discussion and Interpretation

1) Cultivation growth -

After the cultivation the growth was observed varyingly in both raw materials in 15-21 days. The comparative nutritional composition of mushrooms cultivated in different substrates is different. These figures show that all parameters (protein, lipid, carbohydrate, fiber and ash) have some extents of difference within mushrooms cultivated in different substrates.

2) Protein content -

From the observation table the Protein content of mushroom in paddy was significantly higher than in wheat straw.

Amount of protein in wood scrap from the graph is 21mg/ml. & protein contain for wheat straw from the graph is 15 mg/ml.

3) Lipid content-

Lipid content of mushrooms was higher in wheat straw than paddy straw. The amount of lipid in grey mushroom in wheat straw was 0.710g that in paddy was1.27g & the amount of lipid in pink mushroom in wheat straw is 2g and that in paddy is0.37g.

4) Crude fiber contain-

The Crude fiber contents of oyster mushroom in grey mushroom in wheat straw was 0.81g that in paddy was0.81g.

The pink mushroom in wheat straw contains 1.260g that in paddy straw was 0.09g

5) Ash content-

Total ash content in grey mushroom on wheat straw was 81.8(g/100) that in paddy is 57.3 (g/100) for pink mushrooms on wheat straw 7.4 (g/100g) and that in paddy 47.3(g/100g).

6) Moisture content-

The moisture contents of grey mushroom on wheat straw was 87.5% that on paddy straw was77.5%. For pink mushrooms on wheat straw was 92% and in paddy straw was 86.11%.

7) Total carbohydrate contain-

Total carbohydrate content of mushroom in wood scrap was (68.6g/100gm) & wheat straw was (66.7g/100gm).

8) Total metabolizable energy -

Using these data the highest metabolizable energy was found in wheat straw was (300 kcal/100g).The protein and lipid content of oyster mushroom on two different raw materials is near about similar to the finding of Sohi& Alam et al. but carbohydrate , fiber & Ash content are different. The metabolizable energy was also different.

Many factors may be involved in the difference of nutritional composition of mushrooms cultivated in different substrates. In present study suggest that oyster mushroom different from each other in nutritional composition although they are of same genus, however each species grow different raw material with same condition but give high protein contain & low Ash content. Hence fruiting bodies of oyster mushrooms can be taking regularly as a protein supplement or as an alternative protein supplement in their diet. The low lipid & high fiber content of the oyster mushroom make it health beneficial food items especially against heart disease and diabetes.

Economics of Oyster mushroom production (10 Kg/day/300 days)

Low cost Investment

Sl. no.	Item	Quantity	Rate (Rs.)	Total (Rs.)
A.	**Capital Investment**			
1.	Thatched House (15' x 25')	1	30,000	30,000
2.	Chaff cutter (Lever type)	1	2000	2,000
3.	Boiler	1	2,000	2,000
4.	Drum	1	1,000	1,000
5.	Spraying systems	1	1,000	1,000
6.	Biomass stove		1,000	1,000
	Total			**37,000**
B.	**Fixed cost**			
1.	Interest on A @ 15%			5,550
2.	Depreciation (Item 1 @ 30%)			9,000
3.	Depreciation (Item 2,3,4,5,& 6 @ 10%)			700
	Total			**15,250**
C.	**Recurring Cost**			
1.	Paddy straw cost + transport	3.5t	7000	24,500
2.	Spawn @ Rs. 40 / No	2000	40	80,000
3.	Polythene bags for bed & packing	25Kg	135	3,375
4.	Labour @ 1 Per day	300	360/day	1,08,000
5.	Others	--		5,000
	Total			**2,20,875**

Total cost of mushroom production / Year (Rs.)

Working expenditure	:	2,20,875
Total fixed cost	:	15,250
Total Cost	**:**	**2,36,125**

Income (Rs.)

By sale of 10Kg/day @ Rs. 135 for300 days	:	4,05,000
Total cost	:	2,36,125
Net Income per year	:	**1,68,875**

Chapter 10

Cultivation Technology of Button Mushrooms

Button mushroom (*Agaricus bisporus*) is the most popular variety, fetches high price, still dominating the Indian and International market. It contributes about 90 per cent of total country's production as against its global share of about 40 per cent. The method of cultivation of mushroom was recorded as early as 300 BC and their international cultivation was started as early as 600 AD in China. However, commercial production of white button mushroom was initiated in the hilly regions of the country (17- 18°C) like Chail (Himachal Pradesh) Kashmir and Ooty (Tamil Nadu).

Mushroom cultivation slowly spread to North western plains of India (seasonal crop during winter). In Rajasthan, production of mushroom started in 1980. Mushrooms are the health food of the world. Analysis of fresh button mushroom show that they contain 90 to 93 per cent moisture, 28 to 42.5 per cent crude protein, 8.3 to 16.2 per cent crude fibre, 9.4 to 14.5 per cent ash, 59.4 per cent carbohydrates and 3.1 per cent fat. Among the minerals 71 mg calcium, 912 mg phosphorous, 106 mg sodium, 8.8 mg iron and 2850 mg potassium (per 100 g dry weight basis) are present. Among the vitamins 8.9 mg thiamine (B1), 3.7 mg riboflavin(B2), 26.5 mg ascorbic acid (C) and 42.5 mg niacin(B3) are also available on 100 g dry w eight basis.

Mushrooms are good as nutritious food for all ages and under all conditions of health. They are rich in good quality proteins with lysine and tryptophan that are normally deficient in cereals. The carbohydrates in the mushrooms are at a level of 4.5 to5.0 per cent but are in the form of glycogen, chitin and hemicellulose instead of starch. The fat contain is as low as 0.3 % but is rich in linoleic acid, an essential fatty acid.

Cholesterol is absent and in its place ergo-sterol is present which gets converted to vitamin D by the human body. Mushrooms are fairly good source of vitamin C and vitamin B complex, particularly thiamine, riboflavin, niacin, biotin and pantothenic acid. Folic acid and vitamin B12 which are absent in most vegetables are present in the mushrooms which also supply a range of valuable minerals especially potassium and iron.

Mushrooms have traditionally been used for medicinal and tonic properties and cosmetic products. Compounds extracted from button mushroom have been reported to have anti-fungal and anti-bacterial properties . The high proteins, sterols, macro-elements and low calorie content make mushroom ideal for prevention of cardiovascular diseases. Thus they are an ideal food ever for patients, old people, pregnant ladies and children.

Therapeutic properties of mushroom include enhancement of macro phase function and host resistance to many bacterial, viral, fungal and parasitic infections, activation of non-specific immune stimulation and reduction of blood cholesterol and glucose levels.

In the last 10-15 years, large numbers of commercial units have been built by the entrepreneurs/farmers throughout the country for export. There are chiefly two types of mushroom growers in India. These are (i) Environment controlled mushroom growers (high cost technology) and (ii) Seasonal mushroom growers (Low cost technology). Both the technologies arerelevant in Indian conditions. Large environment controlled units grow mushroom mainly for export purpose only while seasonal growers take a few crops of mushrooms in winter months/all year round in hills. There are scores of spawn producing laborites supplying mushroom seed/spawn to mushroom grower in India.

Mushroom cultivation has now become a household name in almost all regions in India. The world's largest button mushroom growing unit is located in India (Punjab). The present production of all types of mushroom in India is about 1,00,000 tonnes in 2008and the bulk of it is the white button mushroom. World's total production of mushroom is about 22 million tonnes with China contributing the lions share.

Mushroom cultivation requires cereal straws for substrate preparation and there is abundant cereal straw available in India for recycling to edible biomass of highly nutritional and medicinal value. India produces nearly 140 million tones of cereals and equal amount of straw is generated by the farmers, which can partly be utilized by the farmers for mushroom cultivation. Land requirement is a minimum and any spare room of the house can be converted into a mushroom growing room, or a hut built on a piece of land can also be used for the purpose.

The raw materials required for crop raising or generated by the farmers on their own fields (paddy/wheat/or any other cereal straw).

The family labour is used for different operations and the only input required from outside is the seed/spawn of mushrooms. This makes the farmer self confident in raising the crop with great remuneration. A farmer grows one crop or two in the season and his income subsisted to a greater extent. The raw materials requirement for raising a crop of mushroom is recyclable cereal straw/organic waste/organic byproducts. The main byproducts used for substrate preparation for mushroom farming are:

a. Wheat / paddy straw.

b. Sugarcane baggage

c. Saw dust

d. Cotton seed meal/soybean meal

e. Scores of the locally available agro byproducts and other agro waste materials like banana pseudo stem/ corn cobs/ groundnut hull etc.

Table1:Economics of button mushroom (*Agaricus bisporus*) cultivation in India under seasonal growing conditions (Low-cost technology system)

A. Expenditure (production costs per crop of 20 tonnes compost)	
1. Fixed cost (facilities used more than one time)	**Rs**
(i) Land rent per crop season (land required-0.5 acres)	2000
(ii) Seasonal growing hut (raw bricks/thatched roof/bamboo) 60×20×10-12 feet (h), 1200 feet @ Rs 25/square foot=Rs 30000 life span of 3 crop seasons, 1 crop season cost	10,000
(iii) Spraying equipments, hoses, tubs, buckets, forks, spades, etc. (life span 5 seasons =Rs 5000), cost for 1 season	1,000
2. Variable costs (materials used one time only)	
(i) Cereal straw – 12 tonnes @ Rs 1000/tone	12,000
(ii) Supplement wheat bran 300 kg/tone – 3600 kg @ Rs 6/kg	21,600
(iii) Urea 20 kg / tone – 240 kg @ Rs 5/kg	1,200
(iv) Cotton seed cake – 60 kg/tone, 720 kg @ Rs 10/kg	7,200
(v) Gypsum – 35 kg/tone, 420 kg @ Rs 2/kg	840
(vi) Spawn- 5 kg/tone, of compost 100 kg @ Rs 50/kg	5,000
(vii) Casing material (decomposed FYM), 2 kg/bag of 10 kg compost-4 tonnes @ Rs 250/tone + Farmalin treatment cost	2,000
(viii) Labour-composting 30 LD, spawning 10 LD casing 10 LD, cropping 60LD, Miscellaneous 10 LD=120 LD @ Rs100/LD	12,000
(ix) Water costs	1,000
(x) Power costs	500
(xi) Pesticides, disinfectants, chemicals	1,000
(xii) Poly-bags for packing	500
(xiii) Poly-bags for cropping	2,000
(xiv) Miscellaneous expenses	2,000
Total (1+2)	81,840

Besides above, poultry manure is also used as nitrogen rich supplement for compost making for button mushroom cultivation. These raw materials are locally available in all the rural areas of the country at very reasonable rates. The farm yard manure (FYM), spent mushroom substrate and coir industry waste are three main ingredients used as casing materials after proper decomposition in button mushroom farming in India.

Bamboo and synthetic cloth are used for making temporary structures (mushroom growing huts), which are also available at very cheap rates all over the country. For seasonal growing no power is required for crop raising excepting the light.

Good quality water which is good for human consumption is good enough for mushroom cultivation as well.

Table 2: Economics of a Plant having 200 tonnes per annum (TPA) production of button mushrooms (High-cost production technology system)

B. Returns
Returns at 12-15% of compost weight in 8 weeks of cropping at an average price Rs 50.00 kg
At 12%, 20 tonnes × 120 kg per ton compost, 2,400 kg Rs. 1,20,000
At 15%, 20 tonnes × 150 kg per ton compost, 3,000 kg Rs. 1,50,000
Profit at 12 % yield, Rs 1,20,000-Rs 82,000 **Rs 38,000**
Profit at 15% yield, Rs 1,50,000-Rs 82,000 **Rs. 68,000**

A. Fixed Expenditure (Assets)Approx. Cost

1. Land procurement and its development

(labelling, plantation, fencing, etc) Rs5,00,000.00

2. Civil works

a)Compost unit: Rs7,00,000.00

b)Production and processing facility Rs33,00000.00

c)Spawn laboratory Rs3,500,00.00

d)Office block Rs2,50,000.00

3. Plant and machinery

a)Compost and casing unit Rs7,00,000.00

(compost turner, filling line, boilers, electrical system, tube well etc)

b)Production unit Rs30,00,000.00

(Chilling plant, low pressure boilers, vapour proof lighting, racks etc)

c)Canning spawn laboratory and other requirement Rs10,00,000.00

(spawn lab. and quality control equipment, canning line etc)

B. Cost of Production

Approx. cost

1. Recurring expenditure

a) Wages and salary: Rs6,00,000.00

b) Expenses on raw materials, energy and fuel Rs14,00,000.00

2. Interest and depreciation

a) On land (15 per cent interest) Rs75,000.00

b) On building (15 % interest and 5 % depreciation) Rs9,20,000.00

c) On machinery (10 % interest and 15 % depreciation) Rs11,75,000.00

Total production at 100 per cent capacity utilization:

At 18 % conversion: 180 tonnes

At 20 % conversion: 200 tonnes

Cost of production per Kg: Rs23.30 (18 % conversion)

Cost of production per Kg: Rs21.00 (20 % conversion)

Profit analysis: At 18 % conversion,

total production: 180 tonnes

Cost of production Rs 41,94,000.00

Sale realization @ Rs 40/kg Rs 72, 00,000.00

Net profit Rs 30,06,000.00

Similarly, at 20% conversion,

Net profit Rs 38,00,000.00

Just after harvest, the mushrooms cannot be stored for more than 24 hours at ambient temperature .

Due to presence of more than 90 per cent moisture content, mushrooms are highly perishable and start deteriorating immediately after harvest. They develop brown colour on the surface of the cap due the enzymatic action of phenol oxidase, this results in shorter shelf life.

Loss of texture, development of off flavour and discolouration results in poor marketable quality and restricts trade of fresh mushrooms.

In view of their high perishable nature, the fresh mushrooms have to be processed to extend their shelf life for off season use. This can be achieved by adopting appropriate post-harvest technology to process surplus mushrooms into novel valueadded products. The value-added products are the need of the hour for the mushroom growers not only to reduce the losses but also to enhance the income by value-addition and boost the consumption of this important horticultural crop.

The possible value-added products can be developed either by converting freshly harvested mushrooms into ketch-up, murabba, candy, chips and pickles or by dehydrating freshly harvesting mushrooms into dehydrated form and then making soup powder, biscuit, nuggets ant RTE.

Chapter 11

Cultivation Techniques of Milky Mushroom

Introduction

- Milky mushroom (Calocybe indica) is suitable for hot and humid climate.
- This variety is new introduction to world mushroom family from India.
- Its high biological efficiency, better keeping quality, simple cultivation process and white attractive colour are factors for its popularity.
- It can be cultivated throughout the year.
- Harvested 24 -28 days after spawning.
- Crop cycle 45 - 50 days.
- Shelf life 3 – 5 days.
- Therefore easy to handling, transportation and storage

Cultivation techniques

- Production of spawn
- Substrate
- Bed preparation
- Spawning and spawn run
- Casing
- Cropping
- Harvesting
- Yield

Flow chart

Straw is chopped in small pieces (2-4 cm size)

↓

Soaked in fresh water for 8-16 hours

↓

Drain out excess water

↓

Straw is submerged in hot water for 40 minutes at 80 – 900C / chemical treatment with Bavistin and formaldehyde / sterilized at 15 lbs for 1 hour Drain of excess water Drying at 60-70 % moist level

↓

Drain of excess water

↓

Drying at 60-70 % moist level

↓

Prepared bed for mushroom

↓

Layer spawning

↓

Spawned bags are shifted to spawn running room and kept in dark (Temperature 25-350C and RH 80% are maintained)

↓

Incubated beds for 20 days (substrate is fully colonized)

↓

The fully spawn run beds can be shifted to cropping room after casing for initiation of buttons.

↓

Take the fully spawn run bed and cut horizontally into two equal halves.

↓

Compact the beds as much as possible by pressing firmly with hand.

↓

Apply casing soil to a height of 1 cm and press it gently.

↓

Spray the water sufficiently to wet the cased soil.

↓

Beds are to be incubated at cropping room (Temperature 30 -350C and RH 80 -90 % are maintained)

↓

Light should be provided for long time (10 -12 hours daily)

↓

Observe the beds daily and spray water, if necessary, to keep the beds wet.

↓

Pin heads appear 8 -10 days after casing

↓

Initiation of fruiting bodies within 3-5days

↓

Harvest the mushroom after 7 days (mushroom 7-8 cm diameter), clean it and pack it in a polythene bag for sales.

↓

Stir the top of the bed after first harvest and spray water regularly. (Second harvest can be obtained in another 10 days)

↓

Disturb the topsoil after second harvest and spray water as regular. (Third harvest can be done after 10 days).

↓

The mushroom yield of 350 –400 g can be obtained from 250 g dry weight of the straw (providing all optimum conditions inside the mushroom shed).

Spawn preparation

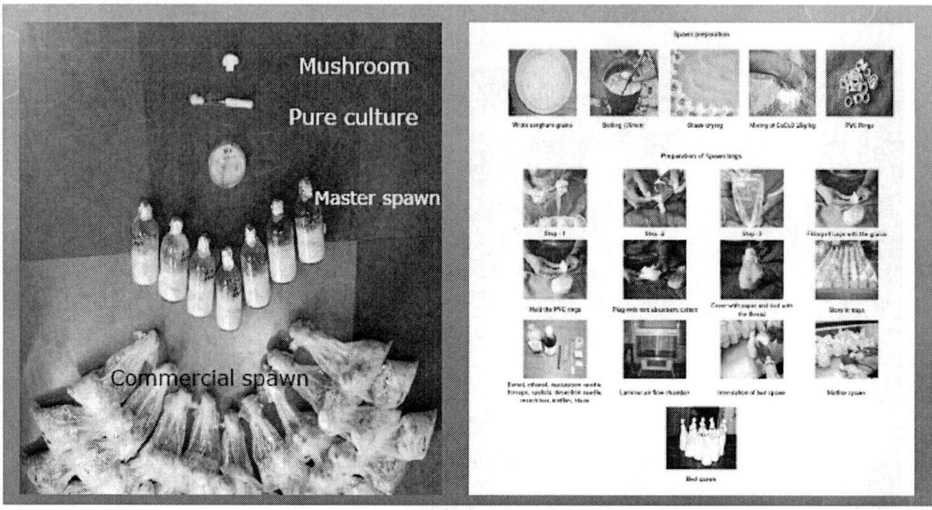

Preparation of beds

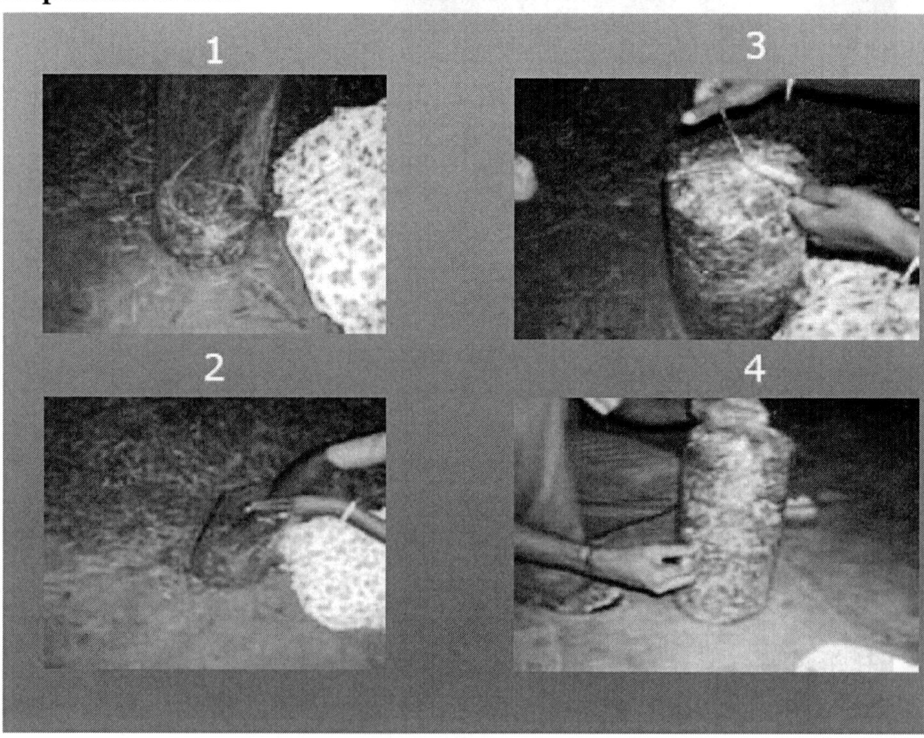

Casing

Harvesting

Economics of Milky mushroom production (10 Kg/day/300 days)
Low cost Investment

Sl. No.	Item	Quantity	Rate (Rs.)	Total (Rs.)
A.	**Capital Investment**			
1.	Thatched House (15'x 20') Blue Poly house- 20'x50' area (1000 sq.ft)	1 1	20,000 60,000	80,000
2.	Chaff cutter (Lever type)	1	2000	2,000
3.	Boiler	1	2,000	2,000
4.	Drum	1	1,000	1,000
5.	Spraying systems	1	1,000	1,000
6.	Biomass store		1,000	1,000
	Total			**87,000**
B.	**Fixed cost**			
1.	Interest on A @ 15 %			13,050
2.	Depreciation (Item 1 @ 10 %)			8,000
3.	Depreciation (Item 2,3,4,5, & 6 @ 10%)			700
	Total			**21,750**
C.	**Recurring Cost**			
1.	Paddy straw cost + transport	3.5 t	7000	24,500
2.	Spawn @ 40 / day	1600	40	64,000
3.	Polythene bags for bed & packing	25Kg	135	3,375
4.	Labour @ 1 per day	300	360/day	1,08,000
5.	Others	--		5,000
	Total			**2,04,875**

Total cost of mushroom production / Year (Rs.)

Working expenditure	:	2,04,875
Total fixed cost	:	21,750
Total Cost	**:**	**2,26,625**

Income (Rs.)

By sale of 10Kg/day @ Rs. 145 for 300 days	:	4,35,000
Total cost	:	2,26,625
Net Income per year	:	**2,08,375**

Chapter 12

Cultivation of Shiitake (*Lentinula Edodes*)

Alternate names: Black Forest Mushroom, Oakwood Mushroom, Chinese Black Mushroom

Origin: Japan, China, Korea

Decomposition: Primary; saprophytic

Description: Brown cap and stem, with white hairs on cap surface. Meaty flavor.

Where to find: Cultivated only; not found in the wild in North America

Ease of cultivation: Easy–medium

Growing season: Spring to fall with plug spawn; year-round with sawdust bags

Growth characteristics: White, cottony mycelium browns as it ages.

Spawn substrates: Grains (rye, millet, popcorn, birdseed); sawdust

Spawn containers: Quart jars to large spawn bags

Spawn incubation temperature: 70–80°F (21–27°C)

Fruiting substrates: Wood: indoors on supplemented sawdust blocks; outdoors on hardwood logs

Fruiting containers: None

Fruiting temperature: 60–70°F (16–21°C)

Fruiting time: 30–60 days from inoculation

Fruiting humidity: 70–85%

Fruiting notes: Mycelium browns as it ages; colonizes in 2–3 weeks. Once full colonization is achieved, expose block to light while still in bag to start fruiting process. Shortly after fruiting begins, mycelium should begin to form an irregular surface, slightly warty, with some warts growing irregularly to the size of cotton balls. When irregular growth is clearly defined, open bag and initiate fruiting conditions. Some growers wait until they see pins begin to form. Pinning process may take 45 days after sawdust is initially spawned.

Harvest: Gently twist mushrooms at base when cap is fully open.

Post-harvest: Cold-shock dunk of Shiitake blocks may be beneficial for subsequent harvests. After first harvest, submerge blocks in cold water for 12–24 hours. Cold shocking stimulates pin formation; submersion in water adds moisture for the fruitbodies in next flush.

Other strains: Many commercial varieties available.

Chapter 13

Growing Shiitake Mushrooms from a Kit

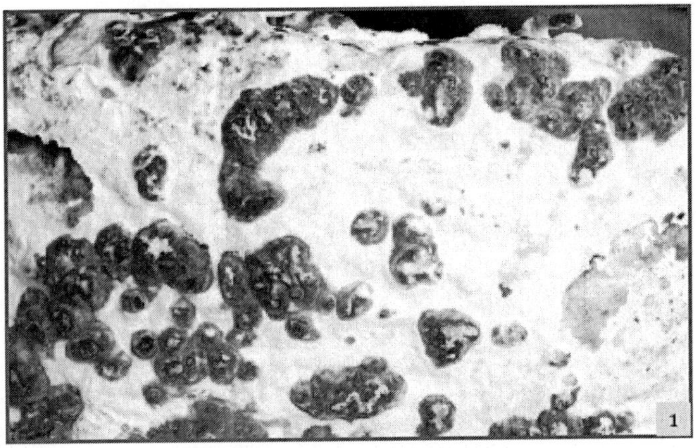

1. **Wait for buttons.** Start by looking at the date on the filter of the plastic bag that contains your block. If it hasn't yet been 45 days since the date listed, do not remove the block from the bag.

Place the bag, with the block still inside and sealed, in an area where it will receive indirect sunlight. Leave it there until 45 days from the listed date, then continue on to the next step. Some people like to leave the block sealed in the bag a little longer, until tiny buttons begin to form on the surface of the block. These are the initial formations of your Shiitake mushrooms. This may take another week or two. If you begin to see these buttons forming before the 45-day period ends, go ahead and continue to the next step.

2. **Remove the block.** Begin by washing your hands. Remove the white block from the plastic bag and place it on a dinner plate. By removing the block from the bag, you're exposing the mushrooms to a sudden influx of fresh air (oxygen) for the first time. This is one of the triggers that encourages the mushrooms to begin fruiting. Another important trigger is the decrease in CO_2 levels.

3. **Assemble a humidity tent.** A humidity tent will keep your block from drying out. Plastic shopping bags are widely available and make excellent humidity tents. Loosely drape the plastic bag over the mushroom block without allowing it to touch the block. One way to support the bag is to stick two or three chopsticks into the block. This action won't harm your block, and will keep the bag off the sides of the block.

4. **Find a good location.** The best location is a spot in the house where the block will not be disturbed by people, animals, or air currents. You also want to put it in a place that will receive some ambient light, but not direct sunlight. Shiitake mushrooms, like many gourmet species, require some light to grow; the mushrooms will not form properly in complete darkness. The optimum temperature for Shiitake is between 55 and 65°F (13 and 18°C). They will grow fine at a normal room temperature in the low 70s, but if you have a cooler location, it may be best to locate your block there.

5. **Mist daily.** To care for your mushrooms as they fruit, all you need to do is remove the humidity tent and mist the block with a common spray bottle. Ideally, this should be done two or three times a day. Once you see mushrooms beginning to form, avoid spraying them directly. Instead, just spray around them. Replace the humidity tent after each misting.

2a

2b

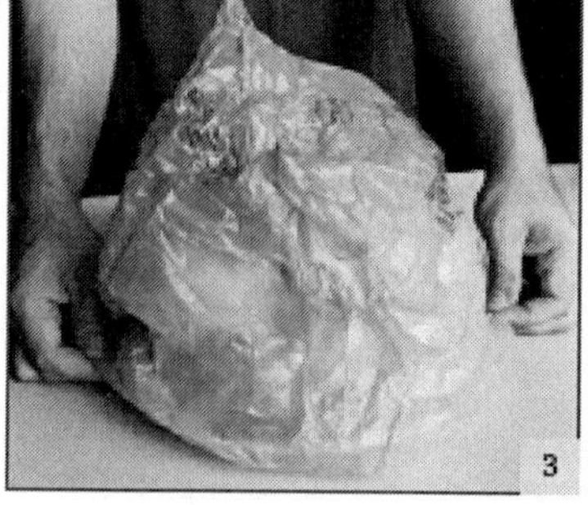

3

6 **Harvest.** Mushrooms will begin to form about 1 or 2 weeks after you remove the original outer plastic bag. The mushrooms will begin as small bumps on the surface of your block, and will soon turn brown and form the typical mushroom stem and cap. You may see the white surface of your block begin to turn into a brown crust as the block begins to age. This is a normal part of the Shiitake growth process.

After a week, soak your block in a pail of cold water overnight. While tap water works fine in most localities, spring water or rainwater is better. Avoid distilled or chlorinated water. The block will float, so you'll have to weigh it down with something to keep it submerged. If any green mold appears on the tough brown skin of the block, just gently rinse it off in the sink. Repeat this soaking process after every flush of mushrooms until the block is no longer producing or is only producing a few mushrooms.

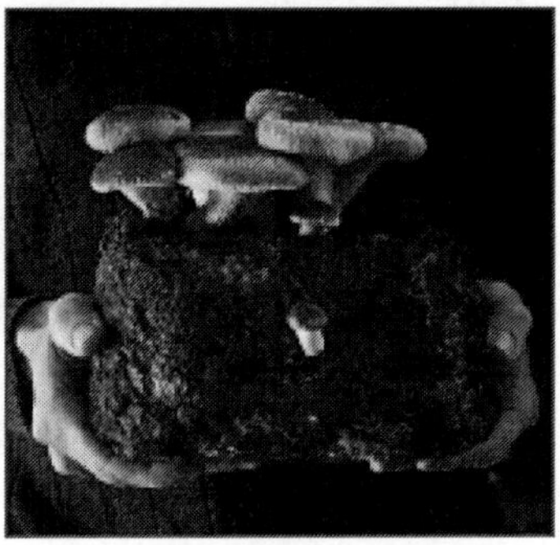

Chapter 14

Growing Shiitakes from Plug Spawn

Plug spawn is the easiest and most common way for people to grow their own mushrooms outdoors. Plug spawn kits, consisting of spiral-grooved dowel pins that have been colonized by mycelium, can be purchased from many mushroom spawn companies. This method only works with wood decomposing species, so you'll be limited to varieties such as Shiitake, Oyster, Maitake, Lion's Mane, and Reishi.

The process is very simple. You provide a log, drill holes into the log, and hammer the wooden dowels into the holes. At this point, the log should be placed in a shady location and allowed to colonize. It generally takes 9 to 12 months before you get your first harvest of mushrooms, but you can expect to harvest mushrooms for 3 to 5 years into the future.

Most species of hardwood logs will work. If you're able to choose your hardwood species, white oak is best. It's a very dense hardwood, which decomposes more slowly than other species. This means that your log will last longer and produce more harvests than softer species of hardwoods.

Other species that will work are maples, elm, sweetgum, other oaks, willow, and alder. Some tree species to avoid are conifers, cedar, hackberry, and dogwood. Look for logs that have a nice layer of bark on them that is not chipped off or otherwise damaged or peeling. The longer the bark stays on the log, the more moisture the log will be able to maintain, and the more harvests you'll get.

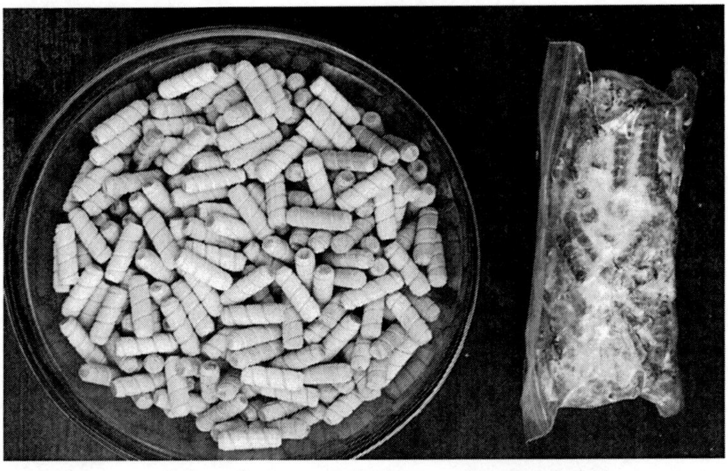

Spiral-grooved dowel pins are the base substrate for plug spawn.

The bark also provides protection from other competitive fungi.

If a log has been allowed to sit out in the open air for an extended period of time, there is a high probability that it will become colonized by some of the many other types of fungi that typically break down and decompose logs in nature.

Older logs are likely to contain these natural fungi, which will compete with the mushrooms you're trying to grow. Try to find logs that have been cut in the last 3 months.

The best time to harvest your logs is when the tree is dormant: anytime after the leaves fall off in the fall, but before the buds start to appear in the spring. The wood contains higher levels of nutrients and sugars at this time, which is advantageous for mushroom growth; however, you can expect success from logs cut at any time throughout the year.

Ideal logs are between 4 and 6 inches (10 and 15 cm) in diameter. If the logs are much thinner than 4 inches (10 cm), they will not hold water as well and will have a tendency to dry out. This inhibits colonization. Alternatively, if you're using logs with diameters greater than 10 or 12 inches (12 or 31 cm), the fungi will have a hard time completely colonizing the logs' interior. Larger logs are also heavier, more difficult to work with, and require many more plugs (which means more cost), without the benefit of many additional mushrooms. Growers generally prefer logs that are between 2 and 4 feet (0.6 and 1.2 m) long.

This method works well for the beginner at home, and it's the same method that's used commercially for Shiitake cultivation outdoors. A home grower may grow three or four logs, while a commercial outdoor Shiitake farm may grow three or four thousand. Regardless of the scale, the process is identical.

Chapter 15

Inoculating a Log with Plug Spaw

1 **Hydrate your logs.** Freshly cut logs should be allowed to sit for around 2 weeks before you inoculate them with plugs. Living trees naturally produce antifungal compounds that begin to dissipate shortly after the tree has died. These compounds will inhibit the growth of your mushrooms, so it's best to wait before you inoculate a freshly cut log.

After your freshly cut logs have sat for 2 weeks, soak them in water for 24 hours before you inoculate them. This will ensure that they have enough moisture for strong mushroom growth.

2 **Prepare inoculation sites.** Using a 5⁄16-inch (0.8 cm) drill bit, begin drilling a straight line of holes, 4 to 6 inches (10 to 15 cm) apart, down the log. These holes should be about 1½ inches (3.8 cm) deep. Use a drill bit stop to quickly and easily get the proper depth. The next line of holes should be 4 to 6 inches (10 to 15 cm) away from the first line, and should form a diamond pattern, with the new holes placed diagonally from the original holes.

3 **Insert plugs.** Inoculate the logs by hammering a plug into each of the drilled holes using a rubber mallet or a common hammer. Ideally, the tops of the plugs should be slightly below the surface of the bark.

4 **Seal the holes.** Seal the holes using melted cheese wax or beeswax. Apply a small amount of melted wax to the top of each inoculation site. This can be done using a brush, a dauber, or any other appropriate tool. The wax seals the drill site, preventing other molds and fungi from competing with your gourmet mushroom strain, and helps keep the inoculation sites from drying out.

5 **Aftercare.** After your logs are inoculated, you need to give them the best colonizing and fruiting conditions possible. Begin by finding a shaded area out of direct sunlight. Under the canopy of the woods is ideal, but any shaded area will work. Mesh shade cloth can help make sunny locations more suitable for inoculated logs. Many growers build stacks with their inoculated logs once they find the right location. If you only have a couple of logs, you can just lean them on the shady side of a barn or house, or lay them on the ground. Some species may need to be partially buried.

In most U.S. climates, normal rainfall should be sufficient to keep your logs hydrated. If there is a dry spell, however, it may be necessary to water or soak your logs again. The goal is to keep them moist. Check on your logs every couple of weeks, and if they seem dry, give them a bit of water.

Many growers cover their logs with straw or a breathable tarp during the winter months. Mold can form if you cover your logs with plastic. Another option is to bring them into the garage for the winter. Keeping them out of the freezing weather will cause less stress on the culture. Soak the logs in water for 12 to 24 hours in the spring to give them a good start for the year.

6 **Wait for fruit.** After your logs have been inoculated and placed in the perfect location, you must wait for them to fruit. If you inoculated them in the spring, you might obtain your first fruits by the fall, depending on such factors as the dimensions of the log, how often you add moisture, the weather, the temperature, etc. It may take up to a full year to get your first fruits, but you should be able to expect repeat flushes for the next several years.

Economics of Spawn Production (100 spawn bags per day)

Sl. No.	Item	Quantity	Rate (Rs.)	Total (Rs.)
A.	**Capital investment**			
1.	Autoclave	1	70,000	70,000
2.	Boiler (GL drum 100 lit. Capacity)	2	2,500	5,000
3.	Culture room with work table (low cost)	1	20,000	20,000
4.	UV lamp with fittings	1	2,500	2,500
5.	Tube light fittings	1	1000	1000
6.	Advance for LPG gas	2	3,000	6,000
7.	Spawn storage room	1	30,000	30,000
8.	Bunsen burner	1	300	300
9.	Heat efficient chulah	1	1000	1000
10.	Glassware & chemicals			5000
	Total			**1,40,800**
B.	**Fixed cost**			
1.	Interest on capital investment @ 15%			21,120
2.	Depreciation (Item 3&7 @ 5%)			2,500
3.	Depreciation (Item 1 2,4,5,8 & 9, 10-10%)			9,080
	Total			**32,700**
C.	**Recurring cost (100 spawn x300 days)**			
1.	Polypropylene bags	150Kg	140	21,000
2.	Cholam grains	8000Kg	26	2,08,000
3.	Calcium carbonate (commercial grade)	160Kg	25	4000
4.	Non-absorbent cotton (400 g rolls)	600	110/roll	66,000
5.	Electricity & Fuel	--	--	60,000
6.	Labour @ 2 men per day for 300 days	300	360/day	2,16,000
7.	Miscellaneous	--	--	10,000
	Total			**5,85,000**

Total cost of Spawn production / Year (Rs)—

Working expenditure	:	5,85,000
Total fixed cost	:	32,700
Total Cost	:	**6,17,700**
Income (Rs.)		
By sale of 30,000 spawn bags @ Rs. 40 per bag	:	12,00,000
Total cost	:	**6,17,700**
Net income per year	:	**5,82,300**

Chapter 16

Cultivation of Lion's Mane (*Hericium Erinaceus*)

Alternate names: Bearded Tooth

Origin: Commonly found on downed hardwood logs across North America

Decomposition: Primary; saprophytic

Description: Large white mass with white spines, beardlike appearance; texture similar to cooked lobster

Where to find: Fallen hardwood logs east of the Mississippi; rarely sold commercially

Ease of cultivation: Easy

Growing season: Late summer/fall

Growth characteristics: Thin, white, wispy mycelium

Spawn substrates: Grains (rye, millet, popcorn, birdseed); sawdust

Spawn containers: Quart jars to large spawn bags

Spawn incubation temperature: 70–75°F (21–24°C)

Fruiting substrates: Wood: indoors on supplemented sawdust blocks; outdoors on hardwood logs

Fruiting containers: Polypropylene bags

Fruiting temperature: 65–75°F (18–24°C)

Fruiting time: 15–30 days from inoculation

Fruiting humidity: 90%

Fruiting notes: Begin by colonizing quarts; spawn these jars into supplemented sawdust bags as the fruiting substrate. After bags colonize for about 2 weeks, cut 2–4 small slits into side of bag. Cut slits even if sawdust block does not appear to be fully colonized, as mycelium of this species can be wispy and does not fill out as thickly as other species. No need to remove colonized sawdust block from spawn bag. Place bag under normal fruiting conditions after slits are made.

Harvest: Twist mushroom off at base when spines begin to elongate.

Post-harvest: Multiple flushes are common; leave substrate in fruiting environment, but cut holes in new locations for the fruits to emerge from.

Chapter 17

Cultivation of Maitake (*Grifola frondosa*)

Alternate names: Hen of the Woods, Sheep's Head

Origin: Commonly found in North America and Japan

Decomposition: Primary, saprophytic

Description: Rosette of grayish petals coming from a common stalk

Where to find: Base of oak trees east of the Mississippi

Ease of cultivation: Difficult

Growing season: Fall

Growth characteristics: Mycelium is cottony and irregular

Spawn substrates: Grains (rye, millet, popcorn, birdseed); sawdust

Spawn containers: Quart jars to large spawn bags

Spawn incubation temperature: 70–75°F (21–24°C)

Fruiting substrates: Wood: indoors on supplemented sawdust blocks; outdoors on buried hardwood logs

Fruiting containers: Polypropylene spawn bags

Fruiting temperature: 50–70°F (10–21°C)

Fruiting time: 30–45 days from inoculation to pinning; another 30–45 days for frond development

Fruiting humidity: 75–85%

Fruiting notes: Likes fruiting temperatures a bit lower than most other species.

After supplemented sawdust spawn bag is fully colonized, leave bag upright and intact for several more weeks, but begin exposing to light. At 30–45 days, grayish clumps of primordial should begin to form on upper surface of block. When clumps are about 2" (5.1 cm) high (which may take 2 weeks), cut a slit in bag just above sawdust block, and introduce fruiting conditions. This allows fresh air to enter, and should encourage frond development.

After 3 days, remove upper half of plastic bag completely, leaving lower half of spawn bag around sawdust block to help retain moisture. Getting to this point may take 60 days or more. Formation of fruitbodies will take an additional 2–3 weeks.

Harvest: When fronds are fully developed and edges begin to darken, remove fruitbody at base.

Post-harvest: Multiple flushes difficult with this species, rarely providing more than one

Chapter18

Wood Ear Mushrooms on 'Sterilized' Substrate

Wood ear mushrooms (*Auricularia spp.*) are commonly cultivated in Asia. Plastic bag cultivation is gaining popularity due to the scarcityof suitable logs and the ease with which different species of Auricularia can be cultivated on sawdust. The technology can be expected to spread in the near future.

There are many Auricularia species of which *Auricularia polytricha*, *Auricularia fuscosuccinea* and *Auricularia auriculu-judea* are the most commonly grown.

1 Substrate preparation

The formula for the sawdust substrate is about the same as for oyster mushroom and shiitake, but the moistening period (fermentation) of the substrate should be longer. The preparation of the bags is again the same.

2 Heat treatment

The filled bags are steamed in the same way as for oyster mushroom and shiitake.

3 Spawning and spawn run

Sawdust spawn is generally used, 10 ml of spawn per bag is sufficient.

During the spawn run the temperature should be 25° to 28° C. The mycelium will cover the substrate in about four weeks.

4 Fruiting

Cuts in the bags are made so the mushrooms can emerge. Take care when handling the bags, because the texture of the substrate will stay soft even after the mycelium has colonised it.

Only little light should be present in the mushroom house. Three to four flushes can be expected. Per bag of 1.2 kg, 300 – 500 g can be harvested.

5 Case description: the Philippines

Although the market in the Philippines is more favourable for the smaller *Auricularia auricula-judae* (the black wood ear mushroom), the temperature range there is more suitable for growing *A.polytricha*. A. *auricula-judea* can only be grown in cooler areas.

Substrate preparation (weight percentages)

- Dry sawdust (moisture content 15-18%) 78 kg
- Fine rice bran (first class) 21 kg
- $CaCO_3$ 1 kg

The rice bran has to be sifted to break bigger particles into smaller pieces. The bigger particles would be the first to become contaminated.

Weigh the substrate ingredients and mix CaCO3 and rice bran well before mixing them with the sawdust. Add water slowly until the moisture content is 65-70%.

Fermentation

Pile the substrate in pyramids and cover with plastic to retain its moisture.

Let the heap ferment for five days and turn the heap on the third day. Sieve through 1.5 mm mesh to remove bigger particles and to break the clumps that may have formed during fermentation. The bigger particles might damage the plastic.

Filling

Pack about 1 kg per 12 x 30 cm bag and add the ring and the plug.

Figure: Putting rings on substrate bags.

Heat treatment

Sterilise the filled bags for 1.5 hours at 121° C or semi-sterilise for 10 hours at a temperature just below 100° C.

Spawning and spawn run

Use one 500 ml of spawn for 50 bags.

The spawn run takes about one month at 25° – 30° C. Place the bags flat in rows on shelves.

A mushroom house (5 m wide, 12 m long and 4 m high) can hold 2640 bags. Each row has

55 bags per layer; four layers per row. Four rows with 220 bags each can hold 880 bags, so that three shelves may hold 2640 bags.

Fruiting and harvesting

The optimum fruiting temperature of wood ear mushrooms (*Auriculariapolytricha)* is 23° – 28° C. To promote primordia formation, the cotton plugs should be removed from the bags and holes cut in the bottom. Try to keep the temperature below 30° C by spraying water and opening the mushroom house at night. The primordia will develop into fruiting bodies in seven to ten days. Twist the fruiting bodies from the substrate by hand, leaving no bits of stem behind.

Chapter 19

The PF Tek Method

This method, originally published in 1995 by an individual who used the pseudonym Psilocybe Fanaticus (PF), involves steaming vermiculite and brown rice flour in small, half-pint canning jars. Once cool, the concoction is injected with a syringe of spores or a liquid culture. Growers ultimately end up with colonized "cakes" of mycelium that are fruited in a small terrarium. PF was arrested in 2005 and prosecuted by the Drug Enforcement Agency for growing illegal hallucinogenic varieties, but he lives on through his legacy—this simple method for home mushroom cultivation.

The materials required for preparing your first PF jars are relatively basic. Brown rice flour and vermiculite will form the substrate.

There are several key reasons why this is a good method for beginners. First, it uses materials that are available everywhere. Vermiculite is available at most garden centers; brown rice flour is available at health food stores and many groceries. Second, it is possible to sterilize this substrate by simply steaming the jars in a pot on the stove. Most of the other methods described in this book require a costly pressure cooker. Third, this method uses spore or liquid culture syringes, commonly available online, as a starting point. Finally, this method only requires a fruiting chamber the size of a small plastic bin. If you're a beginner, I encourage you not to prepare more than 24 PF jars until you've fruited at least one smaller round of jars successfully. Trying to overproduce before you understand all the methods only increases your chances of failure. Start with something small and manageable, and you will be rewarded.

Using the PF Tek Method

Materials

- ❖ Half-pint wide-mouth canning jars
- ❖ Vermiculite (or sawdust, for wood-loving species)
- ❖ Brown rice flour
- ❖ Water
- ❖ Measuring cups
- ❖ Mixing bowls
- ❖ Paper towels
- ❖ Alcohol & swabs or cotton balls
- ❖ ⅛" (3 mm) nail
- ❖ Hammer
- ❖ Masking tape
- ❖ Aluminum foil
- ❖ Large pot with lid or pressure cooker
- ❖ Spore or liquid culture syringe
- ❖ Bleach solution
- ❖ Lighter

PF Tek Fruiting Formula (per jar)

- ❖ ½ cup (8 g) vermiculite or sawdust
- ❖ ¼ cup (60 mL) water
- ❖ ¼ cup (45–50 g) brown rice flour

Time Required

Prep: 30 minutes–1 hour

Sterilization: 1½ hours

Inoculation: 20 minutes

Incubation: 2 weeks

Dunking: 24 hours

Fruiting: 2 weeks

Substrate Preparation for 1 Dozen Jars

1 **Prepare the jars.** Hammer four holes through each lid, using a ⅛" (3 mm) nail. Wipe the interior of each jar with alcohol.

2 **Mix the substrate.** In a large mixing bowl, combine 6 cups (96 g) of vermiculite or sawdust and 3 cups (.75 L) of water. Mix thoroughly. Add 3 cups (500–600 g) of brown rice flour to the vermiculite mixture and mix thoroughly. You should now have enough substrate to fill about 12 jars.

3 **Fill the jars.** Fill each jar with the vermiculite mixture to the level of the lowest ring band. Do not pack it down. Wipe the tops of the jars clean with a paper towel. Fill each jar to the top with a layer of dry vermiculite. This serves as a contamination barrier.

4 **Place the lids.** Place the lid and screw on the ring band. Tape over the holes using masking tape. Fold the tape over to make a small tab — this makes it easy to pull off after it is sterilized.

5 **Sterilize the jars.** Cover each jar with foil. I usually make 8" × 8" (20 cm × 20 cm) squares of foil, then press them on. To steam the jars, put 2" (5 cm) of water in a large pot. Add a small rack or other raised surface inside the pot to keep the jars out of the water, and place the jars on the rack. Put on the cover and steam the jars for 60–90 minutes. If using a pressure cooker, leave the jars in for 45 minutes at 15 pounds per square inch (psi). Let the jars cool in the pot. Don't open it until you're ready to inoculate them. Allow your jars to cool completely to room temperature before inoculation.

6 **Inoculate the substrate.** Move the jars into your glovebox or in front of your flow hood. Remove the foil from the jars. Put your syringes, bleach solution, paper towels, alcohol swabs, and lighter into your glove box.

Spray a piece of paper towel with the bleach solution and use it to wipe down the sides and bottom of the box, the syringes, and any other tools.

Shake the syringe to distribute the spores or culture evenly within the fluid. Remove the plastic needle cover and flame sterilize the needle. If your syringe arrives disassembled, you may need to attach the needle before you can sterilize it. Sterilize by holding a flame to the needle until you can see that it is red hot. Let the needle cool, or fast-cool it by wiping it down with an alcohol swab.

Pull back the tape and inject about ¼ cc of fluid from the syringe into each hole of each jar lid. Cover the hole again after each injection. Repeat this process for all jars.

7 **Incubate.** Place your inoculated jars in a warm, dry, dark place for colonization, or make an incubator for this process (see the next section). For most species, you'll want the temperature of the incubation area to be in the 75–85°F (24–29°C) range.

Let the jars sit and incubate for 2 weeks. You should see initial growth within your jars in 3 to 5 days, although it can take up to 7 days or more. Full colonization of the jars will usually take 2 to 3 weeks. See page 63 for more information on incubation. If you see growth in any of the jars that is any color other than white, remove those jars and dispose of them immediately. Common colors are green, pink, or black. Never open contaminated jars indoors. See page 45 for more information on contaminants.

8 **Dunking.** Once your jars are fully colonized and ready to be placed in the fruiting chamber, you should dunk them in water for 24 hours before proceeding.

Dunking gives the jars a burst of extra moisture before the fruiting process begins, and results in better yields. Working over a sink, simply unscrew the lid of each jar, fill the jar with water, and screw the lid on again.

The colonized cake will probably float, so when you put the cap back on, you may need to force it back down into the water, causing some of the water to overflow. Store the jars in the refrigerator for this 24-hour dunking period. (See the dunking section in FAQ for PF Tek, on page 61, for more information.)

9 **Fruiting.** Place the jars into a fruiting chamber.You may want to consider doubleend casing your cakesto improve water retention for the fruiting process.

Chapter 20

Incubation and Incubators

Many mushroom species have an optimum temperature for their vegetative growth cycle that is slightly higher than 70–75°F (21–24°C). Thus, if your goal is fast growth, your colonizing jars should be stored in a place that has this ideal temperature. Any place in your home will work if it maintains a slightly higher temperature than the average in the rest of your house and is warm, dry, clean, and free of mold or other potential sources of contamination.

If you can't find a place in your home with a consistent ideal temperature, you can make an incubator. Shortly after I started growing mushrooms, We made an incubator similar to the one described here and used it for a few months. Since then, I have simply let my jars colonize on a shelf in the lab room at room temperature, so building an incubator is entirely optional. If your gut tells you this is something you need, or if you'd like to experiment, go ahead. Incubators are not terribly expensive and will most likely save you a day or two in colonization time. It's up to you whether that day or two is worth it.

Make an Incubator

Materials

- 2 heavy-duty plastic bins
- Fully submersible fish-tank heater
- Zip ties
- Water
- Towel or cloth

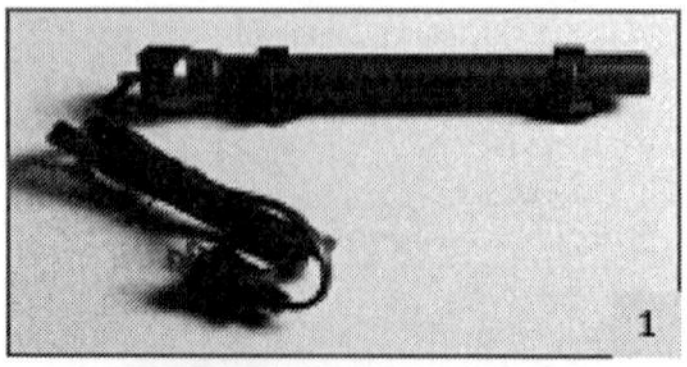

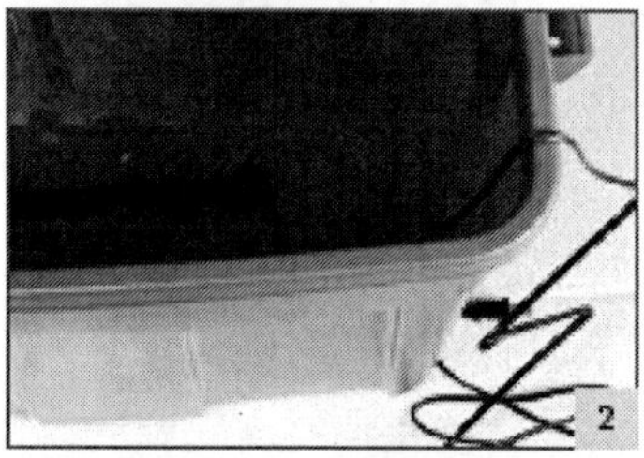

1 **Know your tools.** Fish-tank heaters are a potential fire hazard, so read all directions before using one. The fish-tank heater should come with suction cups that allow it to attach to a surface.

2 **Set up the heater.** Attach the heater to the bottom of one of your bins. Add water to fill the bin several inches above the top of the fish-tank heater. Turn on the heater, and allow it to warm the water to a few degrees above your desired colonization temperature. Once your second bin is attached, you'll no longer be able to make temperature adjustments without removing it.

3 **Place the second bin into the first bin.**

Force it down so the water level rises up the sides in between the two bins. Do not let the second bin touch the heater. Zip-tie the two bins together. Add water until the water reaches at least 6" (15 cm) up the side of the bins. Cover the bin opening with a towel or other cloth.

An Alternate Incubator

This is the perfect option if you want to store your incubating jars. You can store them in a file cabinet, cardboard box, entertainment center, speaker, cooler, etc. Simply open a hole the size of your heater in the lid of a quart jar, fill the jar with water, and insert a 20- to 50-watt fishtank heater into the jar.

Double-End Casing Method

Materials

- ❖ Colonized cakes
- ❖ Several cups of vermiculite
- ❖ Water
- ❖ Spoon

1 **Create the base.** Sterilize your vermiculite if necessarybefore you begin. Place vermiculite in a small bowl. Moisten the vermiculite to field capacity. (The vermiculite should not drip water unless you give it a light squeeze.) Remove the lid of your colonized jar and set it on the counter. This will serve as the base of your mushroom cakes in the fruiting chamber.

Spoon enough sterile, moistened vermiculite from the bowl to fill the jar lid.

2 **Remove the cake from the jar.** If the cake is sticking to the sides of the jar, give it a light tap on the counter to break it free. Be careful not to damage or otherwise break apart the cake itself. Remove the vermiculite layer from your colonized cake in the jar and discard it. You only want to remove the vermiculite layer. If the mycelium has grown into the vermiculite layer, it's okay to remove that mycelium from the cake.

3 **Set the cake on the fresh layer of vermiculite.** Spoon more vermiculite from the bowl onto the top of the cake.

4 **Stack the cakes.** If you're stacking your cakes, place another colonized cake on this layer, and top it off with a final layer of vermiculite.

Fruiting Your Cakes

After your fully colonized PF cakes have been dunked and double-end cased, they're ready to be placed in a *fruiting chamber*, also known as a *terrarium*. A fruiting chamber will give your mushrooms the best possible conditions for fruiting correctly.

Small Fruiting Chambers

There are many types of fruiting chambers out there, and some are fancier than others. I would strongly encourage you to shy away from fancy, elaborate chambers, as they won't work any better than the simpler ones. Until you understand the fundamentals of mushroom cultivation and how its many variables affect each other, I recommend that you keep it simple. Don't introduce unnecessary variables; you'll only have to work through all of them to diagnose contamination problems if you have an issue. Simpler setups always have less chance of failure, are easier to manage, and will make your life more efficient and more pleasurable.

Keep in mind that the suggestions in this chapter are for small-scale fruiting PF terrariums, and are not ideal for largerscale setups. Suggestions for larger setups are discussed in later chapters.

This fruiting chamber is set up and ready to receive cakes.

Air Bubbler

The best addition to perlite in a small fruiting chamber is a fish-tank air bubbler. This consists of a fish-tank pump, plugged into a wall outside the chamber, from which a hose runs to an air bubbler in a glass of water in the bottom of the chamber. The glass of water is filled onehalf to three-quarters full. Using the air bubbler in combination with perlite will not only increase the humidity slightly, but will also continuously add fresh air. a couple of sprays to keep them moist. It's also fine to mist over your cakes gently from well above them, but don't blast them with a strong jet of water. If you spray from above, let the atomized water droplets fall gently to the surface of your cakes.

Humidity Levels

Use the spray bottle on an as-needed basis, in combination with the perlite and the air bubbler, to attain the proper humidity in a small terrarium. For PF cakes, keep the humidity in your fruiting chamber above 90 percent. At this point, humidity gauges or other monitoring are not necessary. If you have several inches of wet perlite in the bottom and an air bubbler, and if you're performing hand misting, your levels should stay near the right point without any other monitoring or tools. If you really want more data, a simple humidity gauge, available at most garden centers or big box stores, will serve that purpose.

Fresh Air

A third important consideration when designing your fruiting chamber is fresh air. As previously mentioned, growing mushroom mycelium produces CO2, which tends to build up within a closed fruiting chamber. Visible symptoms of high CO2 levels are skinny, elongated stems and/ or small caps on the growing mushrooms. The easiest way to ensure that CO2 levels do not get too high is to manually fan your chamber two or three times a day. Just use a scrap piece of cardboard or the

Air bubbler

Spray Bottle

Any standard spray bottle that shoots a fine mist rather than a strong jet of water will work fine. Be sure to avoid bottles that formerly contained chemicals or other toxic substances.

It's best to air out your fruiting chamber two or three times a day. Whenever you air out the chamber, give it three or four mists with your spray bottle as well; however, don't mist your cakes or fruiting mushrooms directly. If the walls of your fruiting chamber are dry, give them lid of your bin as a fan to force fresh air down to the level of the cakes. Fanning semi-vigorously for 5 to 10 seconds, two to three times a day, should do the trick. If you have a fairly small fruiting chamber relative to the size of your mycelia mass, I would increase the number of fannings to four or five times a day.

Please remember that these suggestions are not chiseled in stone. If you miss several fannings, it's not the end of the world, and your mushrooms will probably still grow fine. I've given you suggestions for what's best in an ideal world, but since we're only hobby growers, other things need to take priority sometimes. Going through this process and growing out your mushrooms several times will give you a feel for what they need, as well as what they like and don't like, so try to accommodate your mushrooms as much as possible, and they'll surely accommodate your needs from time to time.

So much more could be said at this point regarding fresh air. For example, there are equations for the ideal amount of air input per unit of volume in the grow room, but these are the concerns of more advanced growers. You don't even need to think about them unless you're considering small-scale commercial production.

Light

Contrary to what most people believe about mushroom cultivation, virtually all varieties require light to stimulate initial growth and to complete the fruiting process of forming mushrooms. The only common species whose cultivation often takes place in darkness is the White Button mushroom. These mushrooms were originally grown in caves in France, but most home cultivators don't take the time to grow them.

The best location for your light source is above the fruiting chamber. Most people choose chambers with opaque sides and a clear top. This prevents a lot of light from the sides, and will generally induce your mushrooms to grow upward.

Mushrooms don't need the high intensity lights needed by green plants. Plants get their energy from photosynthesis; mushrooms get theirs by decomposing organic material. Everyday lights of a normal intensity work fine. For example, a 40-watt bulb is more than sufficient for a chamber holding a dozen cakes or less.

For slightly larger terrariums, hang a shop light overhead. Depending on the size of your terrarium, you might get by with tap lights. Some people use rope lights; others use only ambient sunlight in the room as the light source. Most mushroom varieties will grow fine with a 12 hours on/12 hours off light cycle, coming from whatever light source you choose.

If you control the four main variables — size, humidity, fresh air, and light — within the proper parameters for the fruiting process, you should have very few problems as you proceed.

Setting Up a Small Fruiting Chamber

Most mushrooms begin to fruit about 2 weeks after they're introduced into the right environment. Some growers use the term "birthing" when they talk about the fruiting process. This simply refers to removing the "child" (cake) from the "womb" (jar). Here is a reliable method for setting up a small fruiting chamber in which you can birth your cakes. Small fruiting chambers like these can also be used to maintain the humidity required for the premade kits and Oyster mushroom toilet paper rolls .

To set up the chamber, soak your perlite in a bowl of water. Drain the water, remove the perlite, and place a 2–3" (5–8 cm) layer on the bottom of your bin. Place a pint (500 mL) glass of water in your bin and run a fish-tank bubbler stone from the pump into the water. Choose a light source and place it near your fruiting chamber. Place your cakes on a lid or other base and put them into the chamber.

Making a Desiccant Chamber

Materials

- ❖ Large plastic bin with a tight-fitting lid and screws inserted at strategic points, to hold the layers of screen
- ❖ DampRid or other desiccant
- ❖ Paper towels
- ❖ Two pieces of metal screening, any size mesh, cut to fit inside the bowl

1 **Start with semi-dry mushrooms.** Fan dry your mushrooms to remove most of the moisture. This will make your desiccant more effective and longer lasting.

2 **Set up the first level.** Using the screen, create two levels in your bowl. The lower level will hold the desiccant; the upper level will hold the mushrooms. Place a piece of paper towel on the lower level and place the desiccant on the paper towel.

3 **Add the second level and lid.** Put the second screen into the container. Place the partially dried mushrooms on the screen. Keep the mushrooms in the chamber until they are cracker dry.

Making a Spore Print

Materials

- Recently harvested mushroom
- Thick aluminum foil or glass slide
- Rubber gloves
- Scalpel
- Small juice glass
- Alcohol swab

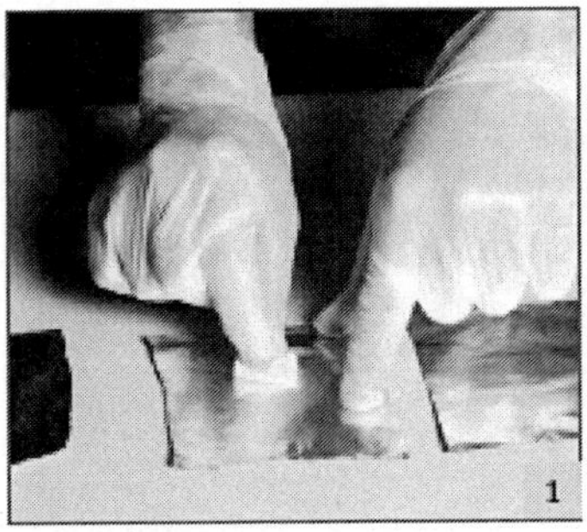

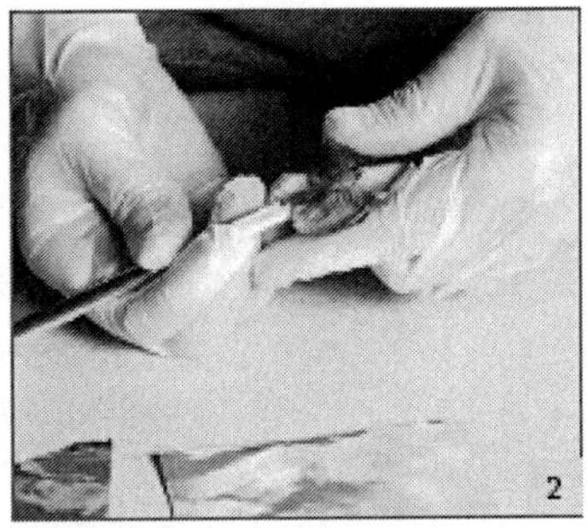

1 **Set up the site.** Find a still-air environment. The inside of a glovebox or a clean cabinet will work nicely. Place the foil or glass into the still-air environment. Wipe the surface down with the alcohol swab. Allow the alcohol to dry for 10 minutes or so with the juice glass over the top.

2 **Harvest the mushroom.** Wearing sterile gloves, harvest a mushroom. Cut off the mushroom cap near the gills, without touching the gills. Place the cap on the foil.

3 **Put the juice glass over the mushroom cap.** This will help prevent contaminants from landing on the spore print. (It helps to place a small, folded piece of paper or something similar under one side of the cup to raise one side a fraction of an inch off the surface. This will prevent a seal from forming and allow some humidity to escape.)

4 **Preserve the spores.** Allow the mushroom cap to sit on the foil for 12–24 hours. The spores will drop from the cap onto your base over this time and stick to the foil or glass. Place your print into a plastic bag with a zip lock. If the print was taken on foil, you may choose to fold the print in half and fold over the sides to seal it up. Label the plastic bag using a permanent marker. Record the species and date.

Making a Spore Syringe

Materials

- ❖ Spore print
- ❖ Glovebox or flow hood
- ❖ Empty 10 or 20 cc syringes
- ❖ Lighter or alcohol lamp

- Two half-pint (250 mL) canning jars
- Scalpel

1 **Sterilize the tools you'll use.** Wrap your syringe, scalpel, and one empty canning jar in aluminum foil. Fill the other canning jar with water. Pressure cook all items, including the jar of water, for 20–30 minutes at 15 psi. If you don't have a pressure cooker, steam sterilize them in a large pot for 45 minutes.

2 **Fill your syringe.** After the items have cooled, move them into your glovebox or in front of your flow hood. Remove the aluminum foil from around your empty jar and syringe. Open your water jar. Draw water from the jar into your syringe. Put the cap back on your syringe.

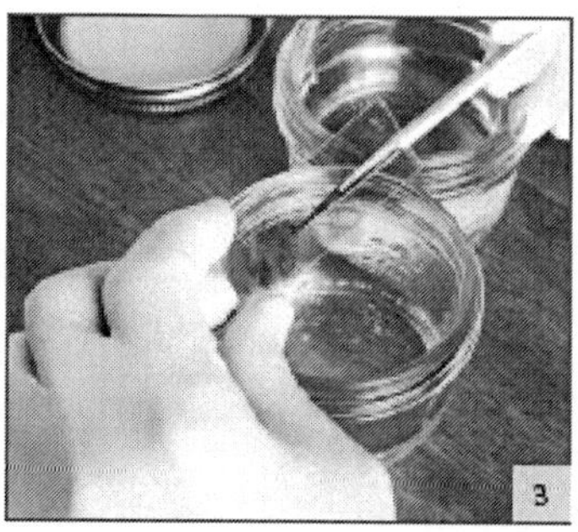

3 **Transfer the spores.** Open up your spore print. Using the scalpel, scrape some of the spores off the print into the empty jar. Most of the spores will be floating on the surface of the water.

A good amount is about one-eighth of the total print. Adjust the amount according to how dark you want your syringes to be.

4 **Mix the spores and water.** Inject water from the syringe into the empty jar. The spores will become suspended in the fluid. Draw the water back into the syringe. You should now be able to see the spores in the water of the syringe. I suggest expelling the fluid from the syringe into the jar and drawing it back up a second time before finalizing the syringe. This will capture more of the spores in the glass and break up clumps of spores. Finally, put the cap back on the syringe. Repeat this process for all syringes you're making. If your prints are more than a week old, let the syringe sit for 12–24 hours before inoculating jars with it. You can use the syringe immediately if the print is fairly fresh.

Chapter 22

Advanced Methods

Agar Culture

Working with agar is one of the most frequently discussed mushroom cultivation methods, but most home growers don't tend to try it until after they've tried many of the other methods. Before the public had access to the Internet, it was very difficult to find sources of inoculum. Now, with liquid cultures and spore syringes so readily and cheaply available, many beginners put off more advanced culturing techniques until they know they'll be sticking with the hobby for a while. But if you've grown mushrooms out completely several times and want to start maintaining your own culture collection, learning the lab methods associated with agar is a must.

What is Agar?

Agar is a gelatinous material derived from seaweed. It's often sold in a powder form, but when mixed with water and allowed to solidify, it turns to a consistency similar to common gelatin. It has no nutritional value by itself, so it must be supplemented with a food source if mushrooms are to grow on it. Some of the most common nutritional additives used to formulate the final agar medium are malt extract and potatoes.

After you prepare this nutritious medium, you must sterilize the mixture in a pressure cooker and pour a thin layer into a container such as a petri dish or small canning jar. When the agar is hot, it flows like a liquid; as it begins to cool, it solidifies into a gel. Your mushroom cultures will grow out on the surface of this gel.

Learning to work with agar efficiently requires a significant amount of time and patience and a certain amount of trial and error. For the best chance of success, I recommend using a flow hood rather than a glovebox. A glovebox restricts your freedom of movement, slowing you down, and generally has much higher contamination rates when used with agar.

Why Use Agar?

Learning to work with agar is important for a number of reasons. The first is that you can use agar to isolate the most vigorous strains from a given species. If you inoculate a petri dish with spores, there may be literally hundreds of different viable strains growing on your dish at once, and not all of them will have the same growth characteristics. As they begin to grow out, the most vigorous and fastest-growing strains tend to dominate.

Working with agar petri dishes allows you to cut away the healthiest sections of growth and transfer them to a new dish, a process called *strain isolation*. After several transfers, you're left with a single vigorous strain of a species that can be tested for other growth factors, such as fruiting time and fruitbody size, allowing you to find the optimal strains for your growing method.

Agar also allows for isolation away from contamination. Even the best mushroom spawn labs experience contamination from other molds or bacteria from time to time, but strains that show signs of contamination can be saved if you'reusing agar petri dishes. You can simply cut away the healthy growth and transfer it to a new dish. It may take several transfers before you achieve contamination-free dishes, but your contaminated cultures need not be thrown away and lost entirely.

A final benefit of working with agar is that you can build a collection of mushroom cultures and maintain it for a long time. Liquid culture syringes remain viable for 6 months to a year when kept in a refrigerator. After that, they need to be grown out and checked for viability.

Agar cultures made in test tubes remain viable for many years if stored properly, potentially even decades. This allows you to build a culture collection of many different mushroom types, and to have each culture ready to grow again instantly if you decide to use it again.

Now that you know the benefits of using agar, you're ready to find out how to use it, including how to prepare media, inoculation, strain selection, cloning, and saving your strains for future use.

Preparing Media

Agar allows you to develop and maintain a collection of cultures that can remain viable for years. They're simply produced with water, agar, and malt sugar.

Within the world of microbiology, there are literally thousands of different types of agar media. Most are specialized mixtures for a specific purpose, such as optimizing or inhibiting the growth of bacteria or fungi. Within the world of mushroom cultivation, home growers tend to focus on two main agar formulations: malt extract agar (MEA) and potato dextrose agar (PDA). These two formulations will be the focus of this chapter.

Agar Containers

Whether you're purchasing or creating your agar media, one of your main considerations is the container that will hold your cultures. Here are a few things to keep in mind. First, the base of your container should not be much larger than 3.5 inches (9 cm) in diameter; a larger container is simply a waste of agar. Second, you'll have to choose between containers that can be sterilized and those that come in sterile packaging. Finally, your container needs a lid. If too much air is allowed to circulate into the dish, you increase the potential for the introduction of contaminants, and you risk drying out the agar and the culture you're trying to propagate. Following are a few of the many types of containers that can work.

Petri Dishes

Petri dishes, well-known in microbiology, are the containers most often used to hold nutrient-rich agar for culturing fungi and bacteria. Petri dishes can be made of glass or plastic.

Glass dishes are made of borosilicate glass, which means they can withstand repeated sterilization and be reused again and again. Most home cultivators use the 100 × 15 mm size. Easy to clean with soap and water, these dishes must also be steam sterilized without agar in them between uses. You can use sterilizer boxes or autoclavable racks to help with sterilization, but instead of purchasing this separate equipment, I stack six or seven empty dishes in a column and wrap them in aluminum foil before pressure cooking for 20 to 30 minutes.

Glass dishes take time to clean and sterilize between uses, but they save you a significant amount of money over time if you make dishes frequently; however, if you're like most home growers, you'll work with agar infrequently, so expensive glass dishes may not be your best solution.

We used plastic disposable dishes for many years with great success. These polystyrene dishes measure 100 × 15 mm, just like the glass ones, and come sealed and sterile in a plastic bag of 20 to 25 dishes.

Plastic containers melt in temperatures commonly attained while pressure cooking, so they cannot be reused, but in all other respects, they're just like glass. There may be times when you open a bag of sterilized plastic petri dishes but only use a portion of the dishes. If this is the case, put the unused dishes back in the original plastic bag and tape it closed. The dishes will remain sterile enough to use in the future.

Canning Jars

Wide-mouth, half-pint canning jars (the same ones used for the PF Tek process discussed in chapter 4) also make good agar containers, and can often be found locally on short notice. Half-pint widemouth jars are shallower than most other sizes, which makes it easier to get a scalpel down to the bottom to cut out segments for transfer.

If you use canning jars when working with agar, don't poke holes into the lid as you do for the PF Tek process. Just turn the lids upside down on the jars so the seal faces upward away from the glass rim. This will allow for a small amount of air exchange but not enough to cause any significant contamination issues.

Glass petri dishes (bottom) are more expensive than plastic ones (top), but they can be sterilized and reused for years. Plastic petri dishes cannot be reused.

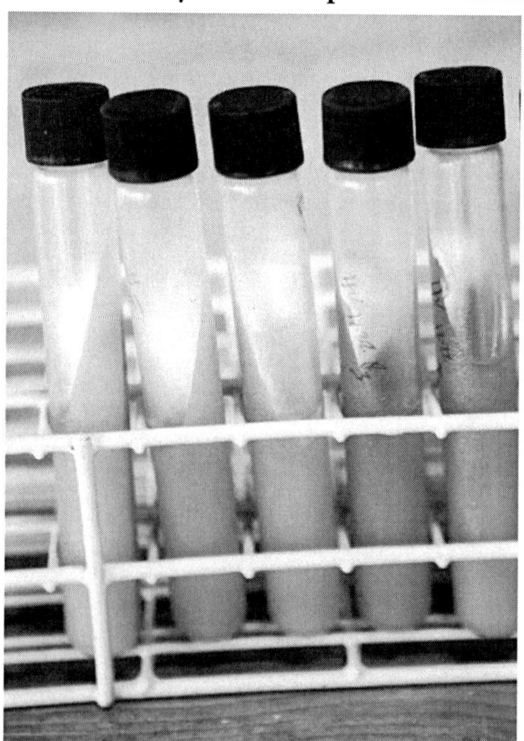

Agar slants are primarily used to store cultures long term.

Test Tubes

A test tube serves several functions when working with agar. Its primary use is as a vessel to store cultures for long periods. The creation of *agar slants*, a process by which the tubes are filled with agar and allowed to cool at an angle, increases the surface area of the agar within a very small space, and allows more room for the culture to grow out than if it were allowed to solidify standing straight up.

Very few cultivators use test tubes as their primary container for common agar work. The mouths of most test tubes are very small, and when you're working to clean up a contaminated culture, it's difficult to cut away the desired section with any amount of precision. There are many other container options, including small plastic drink cups or baby food jars.

Malt Extract

Malt extract in most of its forms, including light malt extract, dry malt extract, barley malt extract, and barley malt sugar, will work for our purposes. Malt extract is available in liquid form, but you need the dry version, which should have the texture of fine household sugar. Look for malt extract in brewing supply or health food stores.

Potato Dextrose Agar

This common formula is used for growing many types of fungi. Potatoes can be purchased in several different forms, and most will work for creating agar. The two methods discussed here involve whole potatoes and instant potato flakes.

Malt Extract Agar

Materials

- 500 mL (2 c.) water
- 10 g (1⅓ tbs.) agar
- 10 g (1⅓ tbs.) malt extract
- 1 g (½ tsp.) yeast (optional)

I usually mix up a large batch of the dry components at one time, and then add 20 g (just over 3 tbs.) of the final mixture into 500 mL (2 c.) of water. So if I pour 200 g (7 oz.) of agar into a bowl, I add 200 g (7 oz.) of malt extract into the bowl as well and then save everything I'm not using at that time. That way, I always have my own premade mix on hand.

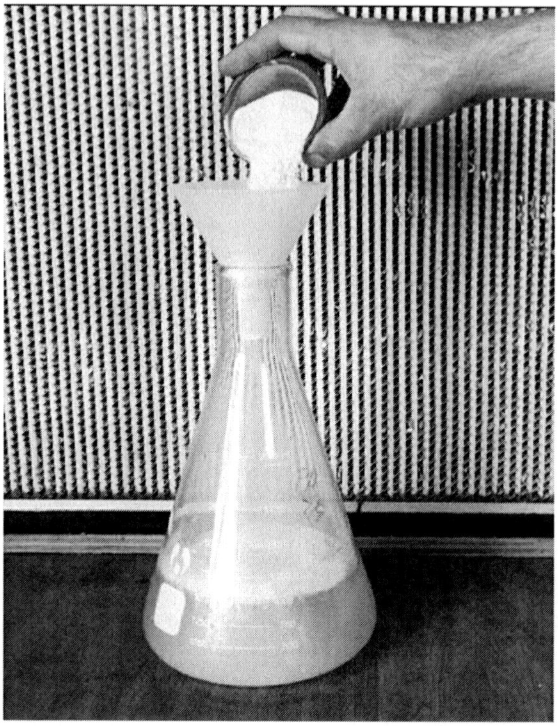

Creating Potato Dextrose Agar

Materials

- ½ to ¾ pound (227–340 g) unpeeled potatoes
- 1 L (4 c.) water, plus additional
- Pot for boiling water
- Cheesecloth or other fine strainer
- 20 g (just over 3 tbs.) agar
- 20 g (just over 3 tbs.) dextrose or dextrose substitute (light corn syrup)
- Pressure cooker
- Flask

1 **Start by creating potato water.** Slice the potatoes to allow more surface area to be in contact with the water; this results in more nutrients in the liquid. Place water and potatoes into a pot and boil them for 30–60 minutes.

1

2a

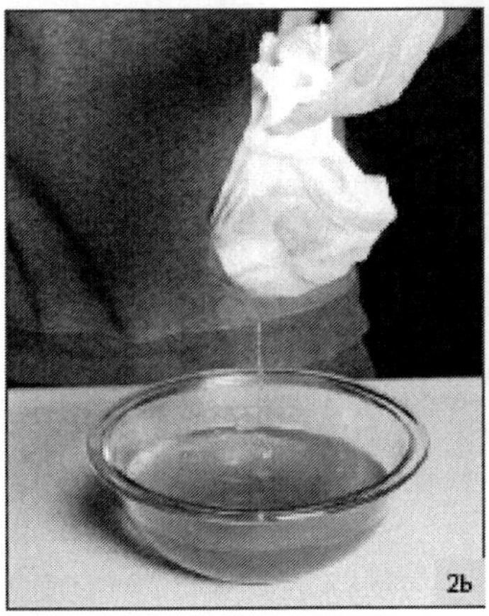
2b

3

2 **Strain the potatoes.** Using cheesecloth or another type of fine mesh strainer, strain the chunks of potato out of the water, but *save the water*. You can throw away the chunks of potato or save them for another use.

3 **Add agar and dextrose.** Add more water to bring the volume back up to 1 L (4 c.). Add agar and dextrose to the water. If you're using a light corn syrup substitute, you need about 4 tsp. Pressure cook the mixture for 30 minutes at 15 psi.

Potato Flake Agar

Potato flakes also work well to create your base of potato water, and this method is even easier than boiling potatoes. You can mix up a large batch of potato flakes and agar, just as you can mix batches of malt extract and agar, and store any extra for later use. Mix all the ingredients together and pressure cook as you would for potato dextrose.

- 500 mL (2 c.) water
- 10 g (1⅓ tbsp.) agar
- 5 g (1½ tbsp.) potato flakes
- 10 mL (2 tsp.) honey or light corn syrup
- 1 g (½ tsp.) nutritional yeast (optional)

Cloning Method

Materials

- ❖ Fresh fruitbody
- ❖ Tweezers
- ❖ Scalpel
- ❖ Petri dish with agar
- ❖ Latex gloves
- ❖ Shot glass or half-pint canning jar (optional)
- ❖ 3% hydrogen peroxide (optional)

Begin by sterilizing your materials. This includes your agar formula and containers (unless you're using plastic dishes that have already been sterilized). Wrap metal tweezers and your scalpel in foil to cook along with the agar.

Prepare your work surface or glovebox. Set the mushrooms you want to clone on

a piece of paper towel. Set your wrapped utensils on another piece of paper towel. When your agar plates have cooled down and solidified, put on your gloves and unwrap your tweezers.

1 **Open the mushroom.** Rip the mushroom you're cloning in half, exposing a section of the mushroom tissue that has not previously been exposed to unsterile air. Remove a small section of mushroom tissue, about the size of a grain of rice or rye, from this sterile section.

2 **Dip the flesh into a jar containing a 10:1 solution of water to hydrogen peroxide.** This is an optional procedure that will help prevent bacterial contamination.

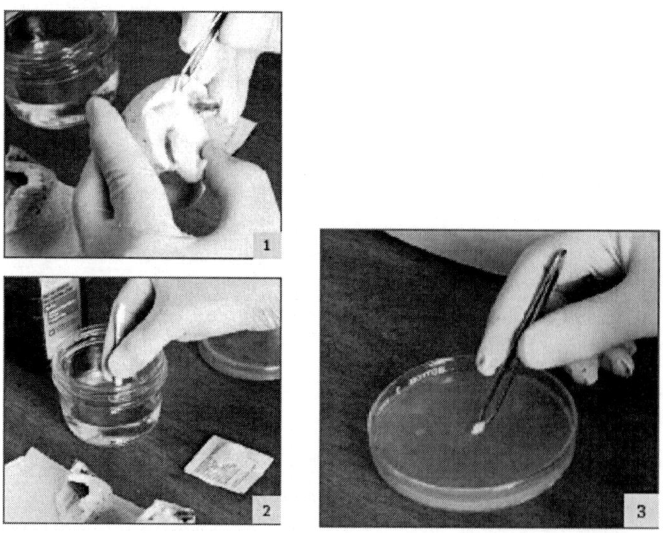

3 **Place this piece of material into the center of the petri dish.** We usually pierce the surface of the agar with the tweezers, and force the tissue down into the agar medium.

This gives the tissue much more surface contact with the agar than if you just set it on top of the medium. The culture will still grow out from the center of the dish, as with any other transfer. Place the lid on the dish. Heat sterilize your tweezers before obtaining another piece of tissue for the next dish. It's usually best to clone 5 to 10 dishes of each species you're working with.

Seal your dishes with Parafilm once all transfers are complete. Label your dishes, noting the species, that it's an initial clone culture, and the date. We also label the dish or make a separate note of the geographic location of the source material.

Bulk Substrates

Bulk substrates are large amounts of any nutritious medium used to grow mushrooms. The term commonly refers to fruiting substrates other than sawdust. Sawdust is technically the most commonly used bulk substrate, but most cultivators who use this term are referring to other media such as straw, manure, or manure-based compost.

Oyster mushrooms, for example, fruit well on sawdust, but most people choose to fruit them on straw because it doesn't have to be sterilized before spawn is applied. Some Agaricus species, such as Portabellas, grow best on manure-based substrates, which don't require sterilization either.

Instead, straw and manure-based substrates rely on pasteurization to make them suitable for mushroom cultivation. Pasteurizing substrates saves a significant amount of time and expense when you're working with large amounts, especially if you're a home grower without access to large commercial pressure cookers.

Pasteurization is a process similar to sterilization, but it requires lower temperatures. Sterilization occurs by attaining 250°F (121°C) for 20 minutes; when pasteurizing, you usually bring the substrate to between 145 and 165°F (63 and 74°C) for 1 hour. The simplest way to explain pasteurization is to say that it kills off the "bad" bacteria and contaminants, and allows the "beneficial" bacteria to survive.

This chapter explains some of the easiest ways to pasteurize small and large amounts of bulk substrates and prepare them for use.

Straw

No special type of straw is required for mushroom cultivation; a bale from any garden center or hardware store will do. Smaller amounts are available at hobby stores that carry mini-bales for craft projects, and can work well if you don't have room to store an entire bale.

Cutting Straw

Straw used as a substrate should ideally be cut into 2- to 3-inch (5 to 8 cm) pieces before being hydrated and pasteurized. Although cutting up the straw is not absolutely required, smaller pieces can be packed more tightly into the container you'll be using, and this will help shorten colonization times. Cut-up straw has more exposed surface area and can potentially offer more nutrients to the mycelium. Also, short pieces of straw are easier to work with.

There are several ways to cut straw, some more labor intensive than others.

If you don't mind a bit of labor, and only need a small amount of straw, you can cut it up, clump by clump, using a large pair of garden shears. You'll have a full pillowcase in 10 or 15 minutes, and it's a good choice if you live in an apartment without much room or outdoor working space.

You can also put straw into a garbage can and cut it up with a weed eater. Just cut a slit in the lid, insert the neck of the weed eater, and start chopping like a madman.

Some people run the straw through a wood chipper. Both of these methods can process a lot more straw but also require outdoor working space.

Straw that's to be used as a substrate should be cut into small pieces. Large garden shears are fine for cutting up small quantities of straw.

A string trimmer can make quick work of chopping straw. Simply cut a hole in the lid of a clean trash can so that the straw stays contained while the trimmer is working.

Hydrating Straw

Like all mushroom substrates, straw must attain the proper moisture level before it can host a growing culture. To ensure the correct moisture level, you should hydrate it before proceeding to pasteurization.

You can either soak your straw in a hot-water bath or steam it. The hot-water bath methods in this chapter will hydrate the straw during the process much more than a steam treatment, and thus require shorter hydration times. If you're using a hot-water bath, soak the straw for 1 to 2 hours before putting it in the bath. If you're steaming the straw, soak it for 12 to 24 hours before implementing the steam cycle. With both of these methods, let all the excess water from the soak drain off before proceeding further.

Manure

The idea of working with manure may sound gross, but let me reassure you, you'll never work with anything that has the smell or texture of fresh manure. In fact, if the manure smells bad or has a disgusting texture, it's too fresh for use in mushroom cultivation.

Mushrooms require manures that are well composted, with the smell and texture of dry dirt. These "aged" manures, whether from horses or cows, have been sitting out in a field or behind a stable for quite a while. Horse and cow manure are basically interchangeable for any mushroom species that grow from this substrate. If you live in the country, anyone near you who has horses or stables probably has more manure than they'll ever use, so don't feel shy about asking for some.

You can also check newspapers in spring or look for ads on Craigslist throughout the year. Garden centers sometimes carry locally sourced, composted manure.

Most garden centers carry branded, bagged manure as well. I've had mixed results with these products, so you may have to do a bit of testing before you find a brand that works well with the species you want to grow. One way to improve the efficacy of bagged manure from garden centers is to mix it 50/50 with another substrate, such as vermiculite.

Horse manure should be well aged before it's used as a substrate

Crumbling Manure

Before it can be used as a substrate, manure must be broken into individual pieces that are smaller than a marble. This is possible only when manure is dry, so any type of manure except the bagged varieties must be dried for several days.

Spread your manure out on newspapers on the porch or in the yard in dry weather, and let the sun dry it. After the manure has dried out enough to crumble easily, it's ready to use.

Ideally, all the individual pieces of manure you'll be using should be smaller than a marble. Break up anything marblesized or larger into smaller chunks. Some people screen their manure based on size, but this isn't really necessary.

Although straw needs to soak before pasteurization, manure does not. It will become appropriately hydrated during the pasteurization process.

Worm Castings

A final type of manure that can work well in many cases is worm castings. There are several brands of castings made primarily from worms digesting cow manure. This is ideal, but most brands don't list the main ingredient they feed the worms. In many cases it's a peat moss base, which will work, but manure-based castings are preferable.

Mushroom Compost

Mushroom compost, available at garden centers, is not a good substrate for growing mushrooms. This spent compost from large-scale Agaricus mushroom farms works fairly well for gardens, but it's not suitable for growing most species of mushrooms. Most of the nutrients have already been extracted by the previous mushroom growth.

Pasteurizing Small Amounts

If you're working with relatively small amounts of substrate, you have several pasteurization methods to choose from: water bath, steam, and oven pasteurization. I'll explain each method and the substrates for which they're most effective.

Water Bath

A water bath is the method I use most often when pasteurizing small amounts of substrate. It works well in the kitchen for preparing small amounts of straw or manure quickly and reliably.

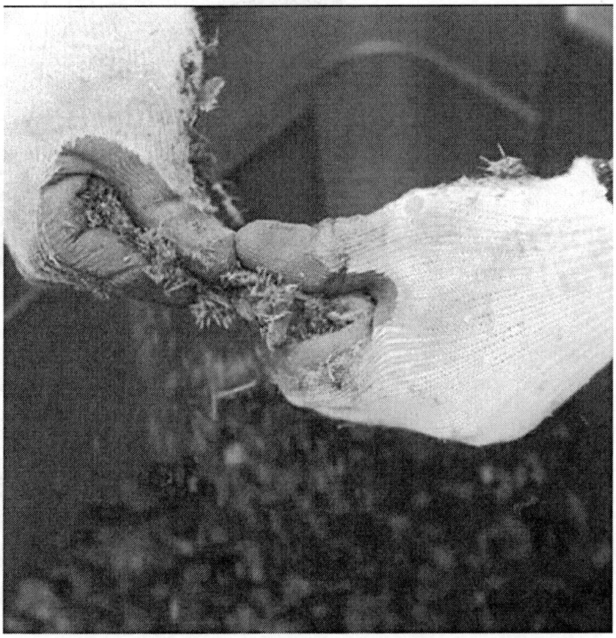

Crumble the manure well before pasteurizing it.

Water Bath Pasteurization

Materials

- ❖ Large pot
- ❖ Standoff plate
- ❖ Pillowcase
- ❖ Thermometer
- ❖ Zip ties (optional)
- ❖ Brick or other heavy object that can be submerged in water
- ❖ Straw or manure

1 **Prepare your substrate.** If you're using straw, cut it into 2" to 3" (5 to 8 cm) pieces , and go to step 2. If you're using manure, crumble it into small pieces , and go to step 3. If you're using straw, soak it for 1 to 2 hours to ensure that it's properly hydrated before proceeding.

2 **Place your substrate into the pillowcase.**

The filled pillowcase should fill the pot at least three-quarters of the way up the side of the pot. Either tie the end of the pillowcase in a knot, or use a zip tie to close it.

3 **Prepare the pot.** Put the standoff plate in the bottom of the pot, and place your pillowcase on it.

4 **Fill the pot with water.** Place the brick or other weight on the pillowcase to keep it submerged.

5 **Heat the water.** Turn on your stove burner and bring the water up to around 155°F (68°C). Keep the temperature between 145 and 165°F (63 and 74°C) for 60 minutes. After 60 minutes, turn off the burner and allow the pot of water to cool.

Once the water is cool enough to touch, pull out the pillowcase and let it finish cooling and draining in the sink for several hours. Once it's completely cool, your substrate is ready to use.

Checking Moisture Content

Before spawning, you need to be sure that the moisture content of your substrate is correct. This is fairly easy with straw, as you can simply let all the water drain off. It becomes slightly more complicated with manure and compost, as they tend to hold different amounts of water, depending on the type or brand, and may remain oversaturated even after draining.

We often use a squeeze test to see if the moisture is right. Gently squeeze a handful of the substrate; only a few drops of water should come out. If there's a stream of water, it's oversaturated, and you need to remove more water before spawning.

For small amounts of manure, you can simply wring it out with your hands. Certain brands of bagged manure or worm castings become compressed and squirt out of your fingers as mud if you squeeze them. It's probably best to consider steam or oven pasteurization for these substrates.

If you give the substrate a firm squeeze and no water comes out, it may be too dry and you may need to add water. This is rarely a concern for substrates pasteurized with a water bath.

Steam Pasteurization

Steam pasteurizing straw or manure is a viable alternative to hydrating it in a water bath. Many large mushroom farms pasteurize substrate using big steam rooms, or they fill shipping containers with substrate and inject steam into them.

Home growers can replicate this process on a smaller scale using a plastic storage bin and a wallpaper steamer. Wallpaper steamers can be found at most home improvement stores and cost between $50 and $100. When selecting a bin, be sure to spend a little extra on one of the thicker models. Steam can compromise the integrity of cheaper bins.

Note that straw should be soaked for 12 to 24 hours and drained before steaming to ensure adequate hydration. Manure and compost should be moistened to near the correct levels before being steamed.

Oven Pasteurization

Cooking manure-based substrates inthe oven is a fast and effective method that tends to work best for prebagged substrates and castings. This method has a couple of advantages. First, it's not very messy — no manure water will be splashed around your kitchen. Second, you don't need any special equipment.

Finally, most of the nutrients are retained within the substrate, unlike the hot-water bath, which allows some nutrients to leach into the water that's is poured off. Note that this method is not recommended for pasteurization of straw.

Steam Pasteurization Method

Materials

- ❖ Thick plastic storage bin
- ❖ Wallpaper steamer
- ❖ Heat-resistant caulk or sealant (optional)
- ❖ Rubber grommet (optional)
- ❖ Manure or straw
- ❖ Compost thermometer
- ❖ Kitchen timer

1 **Cut a hole into the side of the storage bin.** The hole should be the same diameter as the hose coming from the steamer. Cut another hole in the side or top of the bin for the thermometer.

2 **Secure the hose.** Remove the large, flat section that spreads the steam onto the wall. The opening on the end of the hose should be exposed. Place the steamer hose just inside the opening of the bin. Seal it with heat-resistant silicone or caulk if the fit isn't very tight. Also consider using a rubber grommet to seal the hose to the bin.

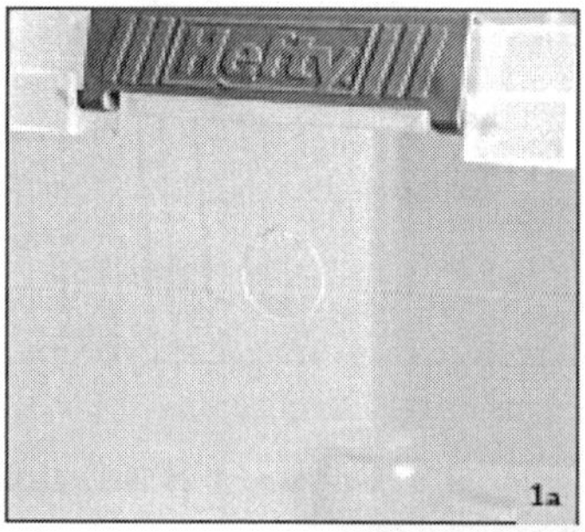

1a

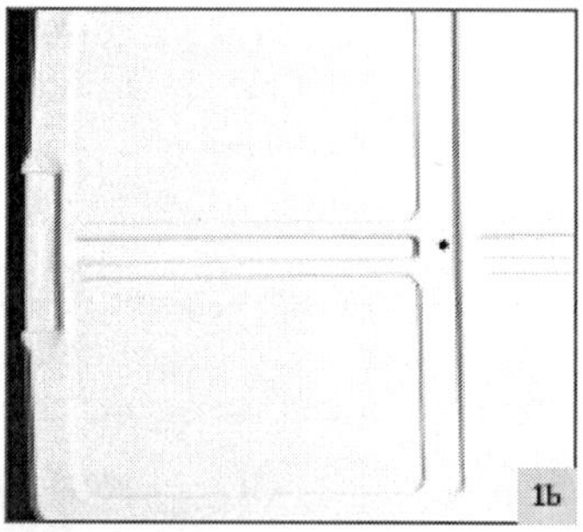
1b

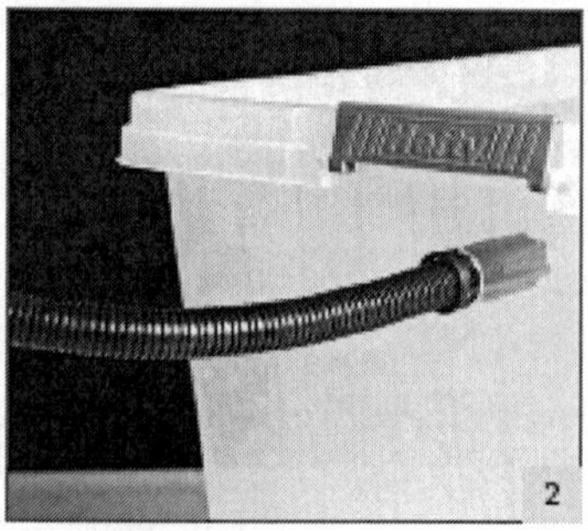

2

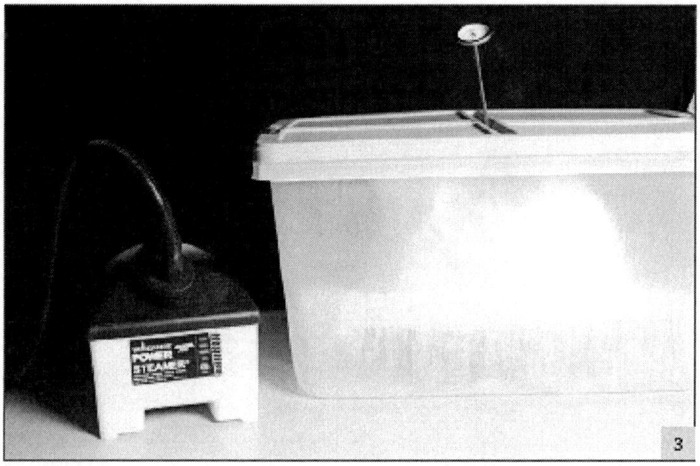

3 **Steam the substrate.** You can use a pillowcase to hold the substrate or putit straight into the bin. Slide the compost thermometer into the bin and into the center of your substrate. Fill the steamer with water and turn it on. Once the temperature on the thermometer reaches 145°F (63°C), set the timer for 1 hour.

If the temperature on the thermometer reaches 160°F (71°C), turn off the steamer until the temperature goes down to 150°F (65.5°C). Turn the steamer back on and repeat until 1 full hour has elapsed within the proper temperature range. When the temperature has stayed at 145 to 150°F (63 to 65.5°C) for 1 full hour, turn off the steamer and allow your substrate to cool in the bin.

Oven Pasteurization Method

Materials

- Aluminum roasting pan, large
- Aluminum foil
- Thermometer
- Water
- Manure-based substrate

1 **Fill the pan with manure.** Prepare your manure by crumbling it appropriately. Wet your manure to just above field capacity (no water should drip from the material unless it's lightly squeezed) and place it in the pan. Ideally, if it's just above field capacity, the evaporation that takes place in the oven will leave it at the right moisture after cooking. Cover the filled pan with aluminum foil.

2 **Heat the oven.** Preheat the oven to the lowest setting, usually between 170 and 250°F (77 and 121°C) for most ovens. Place your thermometer through the foil and into the substrate. Put the pan in the oven. Once the thermometer reads 145°F (63°C), set a timer for 1 hour. Once the thermometer reads 160°F (71°C), turn the oven off. Turn the oven back on once the temperature goes down to 150°F (65.5°C). Repeat this process until one

hour has elapsed at the proper temperature. Turn off the oven once the time has elapsed. Allow your substrate to cool in the oven.

Turkey-roasting pans are a great vessel for oven pasteurizing manure-based substrates.

Preparing the Substrate

After you pasteurize your straw , let it cool down completely before proceeding. When the straw is no longer warm to the touch, it's time to mix the spawn into the substrate and create your column. First, break up the spawn in the spawn bag as much as possible, leaving no chunks or clumps. The more you break it up, the more substrate you'll be able to inoculate with the spawn.

Different growers take different approaches to the next phase. Some growers like to lay the pasteurized straw out on a table, and spread the spawn over the straw before packing it into the tubes. This method probably makes the best use of spawn, as you can spread out all of your spawn fairly equally across the amount of available straw. I prefer a slightly different method: I place a handful of straw into the tube, followed by a handful of spawn, and repeat this process until the column is full. Either way will work fine, so choose what works best for your situation.

Sawdust spawn is a great choice for inoculating straw. Crumble the spawn over the pasteurized straw before loading it into the columns.

Chapter 23

Edible Mushrooms as Functional Ingredients for Development of Healthier and More Sustainable Muscle Foods

Introduction

Muscle foods, such as meat and fish, play an important role in the daily diet of most consumers due to their desirable sensorial attributes and beneficial nutritional properties, including high levels of good quality proteins, vitamins, and minerals. However, muscle foods are deficient in vitamin C, calcium, dietary fibre, and antioxidants . Moreover, consumption of processed food has been linked to certain chronic health problems, such as an increase in diabetes and obesity . The increased awareness of consumers about the possible links between diet and health is leading to shifts in their dietary patterns towards healthier food products. Healthier eating habits include reducing the consumption of ingredients that may cause health problems such as saturated fat, sugar, and salt, and increasing the consumption of ingredients that may promote human health such as unsaturated fatty acids, vitamins, minerals, and nutraceuticals . Adopting these dietary habits favours the maintenance of a healthy weight, as well as minimizing the risk of some lifestyle diseases . As a result, there is increasing demand for healthier food products that consumers can easily incorporate into their diets. Moreover, the raising of animals to produce muscle foods is undesirable from an environmental viewpoint, since it leads to more greenhouse gas emissions, land use, water use, and pollution than growing arable crops .

Consequently, it is also desirable to reduce the total amount of animal foods within the human diet. This can be achieved by avoiding animal products altogether (vegan), avoiding meat products (vegetarian), or reducing the amount of meat products in the diet (flexitarian). This latter approach is suitable for those who want to adopt a healthier and more sustainable diet, but still want to consume some meat.

As a result of these concerns, the food industry is reformulating existing products and creating new products to make them healthier and more sustainable . In this article, we focus on the creation of foods designed for the flexitarian market. Specifically, we focus on replacing part of meat or fish products with healthy and more sustainable natural ingredients: mushrooms. Edible mushrooms are considered to be healthy food ingredients because they contain high levels of quality proteins, dietary fibres, vitamins, minerals, and phenolic compounds . Moreover, they have a relatively low concentration of fat and digestible carbohydrates, which makes them suitable for improving the nutritional profile of foods . Some mushrooms have also been reported to contain constituents that exhibit beneficial therapeutic effects . For instance, polysaccharide-protein complexes and lectins have been reported to have immunomodulatory and antitumor activities , hypotensive effects , and anti-angiogenesis effects . There is, therefore, growing interest in incorporating mushrooms into muscle foods, thereby reducing the proportion of meat present . One of the advantages of using mushrooms for this purpose is that they have good compatibility with meat products because of their umami flavour and fibrous meat-like texture .

This review outlines the utilization of the edible parts of mushrooms as functional ingredients in muscle food products, such as meat and fish. In particular, the impact of mushrooms on the nutritional and quality attributes of these products are critically reviewed, including their physicochemical properties, microbiological stability, chemical stability, and sensory aspects.

Mushroom—A Culinary Delicacy

Mushrooms are valued around the word as culinary delicacies and are popularly known as "vegetable meat" in many cultures. Botanically, they are the fruiting bodies of macroscopic filamentous saprophytic fungi that grow above ground. Their beneficial effects on human health and nutrition were recognized in early Greek, Egyptian, Roman, and Chinese civilizations . Mushrooms can be conveniently categorized into three major groups according to their applications edible (54%), medicinal (38%), and wild (8%) . It has been estimated that there are at least 12,000 mushroom species worldwide, with around 2000 of them being suitable for edible and/or medicinal application, but only 35 being currently cultivated commercially .

Nutritionally, mushrooms have many positive benefits for the human diet: they are low in fat, high in protein, and high in dietary fibre, as well as being good sources of vitamins, minerals, and nutraceuticals . As a result, the global mushroom market has grown considerably over the past few years with 34 billion kg production and per capita consumption exceeding 4.7 kg in 2013 . Indeed, mushroom production is currently a multibillion-dollar industry with an annual turnover of around USD 35 billion in 2015 and estimated to exceed USD 59 billion in 2021, growing at around 9.2% from 2016 to 2021.

Commercially, mushrooms are mainly cultivated on agricultural residues, which enables these waste materials to be converted into a valuable human food source , while also reducing waste and environmental pollution.

Some of the most important commercially cultivated mushrooms are Agaricus bisporus (agaric or button), Lentinula edodes (shiitake), Flammulina velutipes (enoki or winter mushroom), Pleurotus eryngii (king trumpet mushroom), Pleurotus ostreatus (oyster mushroom), Volvariella volvacea (paddy straw mushroom), Calocybe indica (milky mushroom), Hericium erinaceus (pom pom or lion's mane mushroom), Boletus edulis (porcini, cèpe, or king bolete mushroom), Grifola frondosa (maitake or hen of the woods mushroom), and Agrocybe aegerita (pioppini) . Notably, about 85% of the world's cultivated edible mushrooms is represented by only five genera viz. Lentinula, Agaricus, Pleurotus, Auricularia and Flammulina , despite the fact that a wide variety of other edible mushrooms could also be cultivated commercially on a large scale. There are some extrinsic and intrinsic factors that influence the stalk height, stalk diameter and cap size in cultivated mushroom. The most important factors responsible for increased production of cultivated edible mushroom are temperature, humidity, fresh air, and compact material.

Mushroom as Bioactive Functional Food Ingredients

Although many ingredients are used while preparing processed food products, the role of food ingredients that merits special mention are those with inherent nutritional as well as functional properties that influence the quality of finished food products. Therefore, ingredients are now considered as an essential part and parcel of any food product development process. However, the ingredients or compounds obtained from natural sources and generally regarded as safe are of great interest because of their safety and health characteristics . As per the Food and Drug Administration, these are the substances that influence various attributes and properties of any food, either directly or indirectly. They are included at any down streaming stage of processing, be it production, packaging or storage of food, till it reaches the consumer. The purpose is not only to improve nutritional quality and safety but also the freshness, appearance, and overall acceptability of the food products by modifying taste and texture. These additives are often considered as nutraceuticals when these or part of their components exert medical or positive health benefits and play a vital role in the prevention and treatment of various diseases.

The nutraceuticals could be either whole food or its part, or even a single component or extract of food which is regularly being used as a dietary supplement. A food item is termed as "Functional" only when these nutraceuticals are incorporated in the food or its formulation to achieve specific target function such as improving the well-being as well as quality of human life by reducing the risk of disease beyond its nutritional value.

Mushrooms, which belong to filamentous higher fungi, are known for their nutritional richness, low caloric value, taste, and nutraceutical properties. Due to their unique nutritional as well as textural properties, they are used as a dietary supplement and often considered as an alternative source of meat, fish, vegetables, fruits, etc. Moreover, mushrooms are a source of high-quality protein produced in huge quantity from recycling worthless agro-wastes including agro-industrial waste per unit area and time.

Therefore enriching or fortifying diets or food products with such a good source of protein containing all the essential amino acids may help in reducing the incidences of proteinenergy malnutrition in humans. Furthermore, owing to the presence of numerous secondary metabolites or nutraceuticals or biologically active compounds having medicinal value, mushrooms can also be used as bio-therapeutic agents.

Generally, mushrooms possess all three functionalities of food—nutrition, taste, and physiological functionalities. Mushrooms have a peculiarly pleasant savory taste called umami due to presence of sodium salts of free amino acids such as glutamic and aspartic amino acids and 50-nucleotides. The umami taste, also called the palatable taste, is nothing but the overall food flavour enhanced by mono-sodium glutamate. Hence, mushrooms are preferable and adaptable in most food formulations due to this unique flavour.

Again, the umami taste peptides and umami-enhancing peptides are also considered to be important components which influence the sensory quality of mushroom. Peptides with different structures and length possess unique taste properties including sweet, bitter, umami, sour and salty. They are usually tasteless in water, but they can increase the salty, sweet, sour, bitter or umami taste in combination with corresponding tastants.

Various researchers have reported that some dipeptides or tripeptides containing Glu such as Glu-Glu, Glu-Asp, Glu-Asp-Glu, Glu-Gly-Ser enhance umami taste . Recently umami peptides (2 tripeptides and 3 dipeptides) were isolated from hydrolysates of dried shiitake mushroom and these peptides are believed to be responsible for specific taste of shiitake mushroom. They also contribute to the unique taste of mushrooms or even interact with other volatile compounds to influence the whole flavour of foods . In an another study, umami taste peptides like Gly-Leu-Pro-Asp and Gly-His-Gly-Asp isolated from the mushroom Agaricus bisporus are reported to act as key molecules for kokumi taste.

Kokumi taste is best described as flavor characteristics such as mouthfulness, complexity, and continuity. Kokumi taste substances have slight taste or even no taste by themselves, but they can enhance the flavor of the basic tastes, such as sweet, salty, and umami. Interestingly, when incorporated to a blank chicken broth, these peptides from Agaricus bisporus can elicit new taste sensations, such as mouthfulness and complexity.

Nutritive Profile of Mushrooms

As stated earlier, mushrooms are excellent sources of dietary fibre and rich in protein possessing all nine amino acids that are essential for humansbut low in fat and calories. In general, the mushroom fruit bodies contain 5–15% dry matter, 19–35% protein and low fat content. Indeed, the protein content of mushrooms is almost four times greater than tomatoes and carrots, six times greater than oranges, and 12 times greater than apples.

Mushrooms, both pilei and stems are excellent source of dietary fibre mainly due to the presence of non-starch polysaccharides. Stems of mushroom are mainly composed of insoluble dietary fibre (IDF) and glucans. Hence, mushroom stem could be utilized for preparation of biologically active polysaccharide complexes as food supplement. In a recent study, stem of enoki or winter mushroom (Flammulina velutipes) reported to contain 32% dietary fibre. Again, the low fat and high-fibre content of mushrooms may help in preventing hypertension and hypercholesterolemia, as well as being beneficial in weight control. Mushrooms are also healthy sources of essential fatty acids (52–87% unsaturated fatty acids), mostly in the form of linoleic acid, which cannot be directly synthesized in the human body but is required for health. Mushrooms are also rich in indigestible carbohydrates, which makes them promising sources of novel prebiotic components. The low glycemic index and high mannitol content of mushrooms is also believed to be beneficial for diabetics.

Being an excellent source of dietary fibres and proteins, mushrooms also have a low sodium content, and contain a diverse range of micronutrients, such as vitamins B1, B2, B12, C, D, E, niacin, and folate. Despite being cultivated in the dark and being a non-animal source of food, mushrooms contain significant levels of vitamin D, which is often referred to as "the sunshine vitamin". Upon exposure to sunlight or ultraviolet (UV)-B light, the vitamin D (specially vitamin D2) content of mushrooms increases appreciably, which can play a significant role in the bone and cartilage health of vegans and vegetarians.

Table 1. Chemical composition of some common and popular mushroom species.

Mushroom	Common Name	Protein	Fat	Crude Fibre	Ash	Carbohy
Fresh/Raw (g/100 g)						
Agaricus bisporus	White button, Agaric, Pizza	3.00	0.34	1.45	0.79	3.69
Flammulina velutipes	Winter, Enoki	2.66	0.28	2.80	0.91	8.42
Grifola frondosa	Maitake	1.94	0.20	2.70	0.52	2.70
Pleurotus ostreatus	Oyster	2.00	0.99	2.10	0.24	5.35
Pleurotus sajor caju	Oyster	23.3	3.0	35.6	3.2	65.5
Dried (g/100 g)						
Pleurotus eryngii	King trumpet oyster	28.8	3.0	-	3.5	52.2
F. velutipes	Winter, Enoki	18.42	2.94	7.81	6.33	56.37
Termitomyces heimii	Wild edible	23.75	3.58		4.40	54.70
A. bisporus	White button, Agaric, Pizza	29.29	2.22	24.56	7.12	20.57
P. sajor caju (stalk)	Oyster	22.51	2.6	16.24	8.54	40.2
P. sajor caju (cap)	Oyster	26.34	3.07	8.97	10.37	38.17
P. ostreatus	Oyster	20.04	8.65	-	7.78	60.21
Tricholoma nauseosum	Matsutake	18.1	2.0	30.1		31.1
Sarcodon imbricatus	Scaly hedgehog	12.0	2.8	5.1		64.6
G. frondosa	Maitake	21.1	3.1	10.1	7.0	58.8
Hericium erinaceus	Pom pom or Lion's mane	22.3	3.5	7.8	9.4	57.0
Boletus aereus	Bronze bolete or The dark cep	17.86	4.4	-	8.87	72.83
Boletus edulis	Cep or Porcini	21.07	2.45	-	5.53	70.95
Boletus reticulatus	Summer cep	22.57	2.55	-	19.72	55.16

F. velutipes (stem waste)	Winter, Enoki	13.50	1.47	32.30	8.24	63.89	-
P. florida	Oyster	27.83	1.54	23.18	9.41	32.08	-
Russula delica	Milk-white	26.25	5.38	15.42	17.92	34.88	-
Lyophyllum decastes	Fried chicken	18.31	2.14	29.02	14.20	34.36	-
Fistulina hepatica	Beefsteak fungus	63.69	2.63	-	11.30	22.98	364
Laccaria laccata	Deceiver or Waxy laccaria	62.78	3.76	-	20.69	12.77	336
Suillus mediterraneesis	-	24.32	2.61	-	27.64	45.42	302
Tricholoma imbricatum	Matsutake	50.45	1.88	-	6.45	41.21	383
Volvariella volvacea	Paddy straw	29.5	5.7	-	10.4	60.0	374
Lentinula edodes	Shiitake	17.5	8.0	-	8.0	67.5	387
Auricularia polytricha	Wood ear, Jelly ear	7.7	0.8	-	14.0	87.6	347
Tremella fuciformis	White Jelly	4.6	0.2	1.4	0.4	94.8	-
Pholiota microspore	Nameko	20.8	4.2	-	6.3	66.7	372
Calvatia utriformis	Mosaic puffball	20.37	1.90	-	17.81	59.92	744
Lycoperdon echinatum	Spiny puffball	23.52	1.22	-	9.43	65.83	544
Russula cyanoxantha	Charcoal burner	16.80	1.52	-	7.03	74.65	590
Agaricus campestris	Field or Meadow	18.57	1.1		23.16	58.16	-
Boletus armeniacus	-	18.25	1.56		12.09	68.10	-
Tricholoma giganteum	Matsutake	16.1	4.3	4.5	5.0	70.1	-
V. volvacea	Paddy straw	30.1	6.4	11.9	12.6	50.90	-

Furthermore, mushrooms are a rich source of essential minerals like iron, copper, manganese,and zinc that play an important role in the proper functioning of different metabolicnpathways. Indeed, the levels of some important trace elements (such as potassium and phosphorus) are typically considerably higher in mushrooms than in most vegetables. Edible mushrooms can also produce a range of flavonoids, which may exhibit health benefits.

It should be noted that the nutritional composition of mushrooms varies considerably depending on factors such as species, intra-species genetic variability, maturity, growth conditions,geographic location, environmental conditions, and post-harvest conditions. The chemical composition of some popular varieties of mushroom reported by different researchers, are presented in Table 1.

Nutraceutical Components in Mushrooms

Mushroom nutraceuticals are natural compounds found in mushrooms that may have health benefits by reducing the risks of certain diseases or by improving human performance. The potential health-promoting and disease-preventing effects of mushroom nutraceuticals have been attributed to a broad range of biological activities, which are discussed in this section.

Mushrooms have been reported to contain different kinds of nutraceuticals, including lectins, triterpenoids, ganoderic acid, _-glucan, phenolics, flavonoids, hispolon, calcaelin, proteoglycan, lentinan, laccase, nucleosides, nucleotides, and ergosterol . As far as polyphenolic compounds, are concerned fruiting bodies of mushrooms as well as mushroom extracts contain significant amounts of phenolic acids, especially derivatives of benzoic acid and derivatives of cinnamic acid. Different mushroom species have been found to contain protocatechuic, p-hydroxybenzoic, vanillic, salicylic, p-coumaric, gallic, gentisic, syringic, veratric, cinnamic, caffeic, and ferulic acids . The biological activities and potential health benefits of some of these nutraceuticals have been extensively studied. For example, a number of polysaccharides found in mushrooms, including chitin, _-glucan, _-glucan, mannans, xylans and galactans, have been reported to have potential health benefits. In general, the nutraceuticals in mushrooms may exhibit a broad spectrum of different biological activities depending on their chemical structure and their interactions with biochemical processes, including anti-inflammatory, anticarcinogenic, antitumor, antimutagenic, antidiabetic, antibacterial, antiviral, anti-obesity, and anti-hypercholesterolemic activities . As their application in promoting human health have been extensively reviewed by many previous researchers, we do not consider them further in this review.

Prebiotic Effects of Mushrooms

There is growing evidence that human health can be promoted by consuming a diet that establishes a diverse microbiome in the colon. In particular, diets that favour the growth of beneficial bacteria such as Lactobacillus and Bifidobacterium, while suppressing the growth of detrimental bacteria such as Clostridia and Bacteroides, may promote health . Prebiotics are non-digestible and fermentable food components, such as oligosaccharides, dietary fibres, and non-digestible starches, that promote health by selectively modulating the composition and/or activity in the intestinal microbiota.

Mushrooms have been reported to contain numerous constituents that exhibit prebiotic activities, including chitin, hemicellulose, _-glucan, _-glucan, mannans, xylans, and galactans . Some of the important mushroom species that have been reported to exhibit strong prebiotic activity include L. edodes (Shiitake), Trametes versicolor (Yunzhi), and Ganoderma lucidum (Reishi). In a study, Chou et al.reported that prebiotics (polysaccharides and proteinpolysaccharide complexes) from mushrooms passed through the human stomach and small intestine without digestion, then reached the colon where they stimulated the growth of healthy bacteria (Lactobaccilus acidophilus and Bifidobacterium longum subsp.). Similarly, glucans from P. ostreatus and P. eryngiiand G. lucidumhave also been shown to stimulate the growth of Bifidobacterium sp. and Lactobacillus sp.

Prebiotic mushroom polysaccharides are also reported to exhibit antiobesity and antidiabetic effects by regulating the energy homeostasis and plasma glucose levels of the host. The restoration of energy balance is believed to be due to the supply of alternate energy sources from short-chain fatty acids produced during fermentation of the non-digestible carbohydrates in the colon . Some in vitro studies have also reported that extracts from G. lucidum can modulate the gut microbiota in a manner that may help prevent obesity. Other studies have reported that the polysaccharides from various mushroom varieties may be able to ameliorate metabolic syndromes (including diabetes), such as Agaricus brasiliensis, Agrocybe chaxingu, Catathelasma ventricosum, Pleurotus abalonus, Tremella fuciformis, G. frondosa, and G. lucidum. The potential health promoting and medicinal properties of various bioactive ingredients found in mushrooms are summarized in Figure 1.

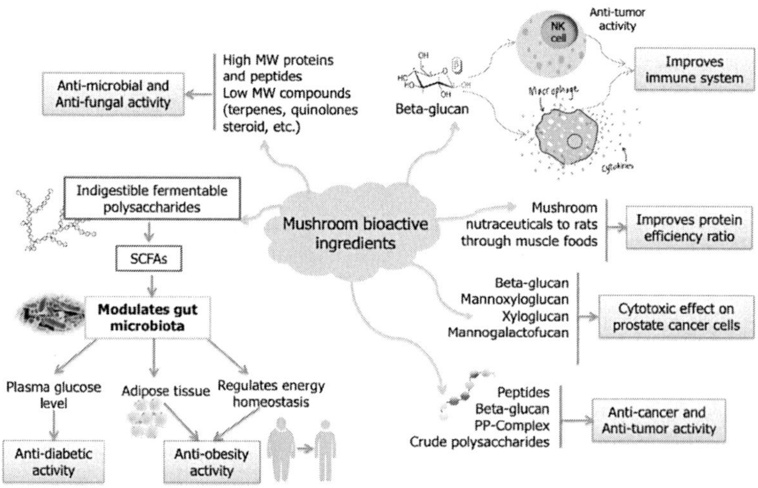

Figure 1. Schematic diagram depicting health promoting and medicinal properties of mushroom bioactive ingredients (MW = Molecular weight; NK cell-Natural killer cell; PP-complex: Protein-polysaccharide complex; SCFA = Short-chain fatty acids).

Effects of Edible Mushrooms on Muscle Food Products

Edible mushrooms, due to richness in nutritive value and functional food components, make them an unmatched source of healthy food and are regarded as superior nutritional supplements. To harness the goodness of nutritional, nutraceutical and other medicinal values, mushrooms are not only used directly as food but also as raw materials in formulation and development of new functional foods for health-conscious consumers.

Other than these values, mushrooms are preferred as additives by the food processors due to their aroma, taste and inherent texture-modifying functional propertieswhich are reported to positively influence the flavour, appearance, overall acceptance and shelf-life, when incorporated in various processed food formulations

Considering their enormous benefits, varieties of food products such as breads , fish and meat products, cookies, other preparations like instant soups, pasta, snack seasonings, casseroles, and rice dishesare being formulated incorporating mushrooms as functional bioactive components that is stated to improve the nutritional profile and potential health benefits. Although quite a large number of research articles are available highlighting the use of mushrooms as potential functional compounds in various food applications, this review limits its focus on the potential application of mushrooms in muscle foods (meat and fish) only.

Incorporation of mushroom and its parts not only influences the desirable texture, taste, flavour, and stability of muscle food products considerably but also enriches them with nutritive and functional health values. Figure 2 indicates the beneficial effect of mushrooms on quality attributes of muscle foods and associated health benefits. Further mushroom is better known for its low sodium content. For example, the fruiting bodies of *Agaricus* sp. contain 396 mg sodium/kgwhich is low amongst the vegetables. On the other hand, processed meats contain 7–39 g sodium chloride/kg. Dietary intake of such a higher amount of sodium is often linked with various diseases and increases the risk of hypertension and cardiovascular diseases. Therefore, pre-mixing or blending mushrooms in processed meats may help in reducing the sodium content of the products, offering more nutritional and health benefits to consumers.

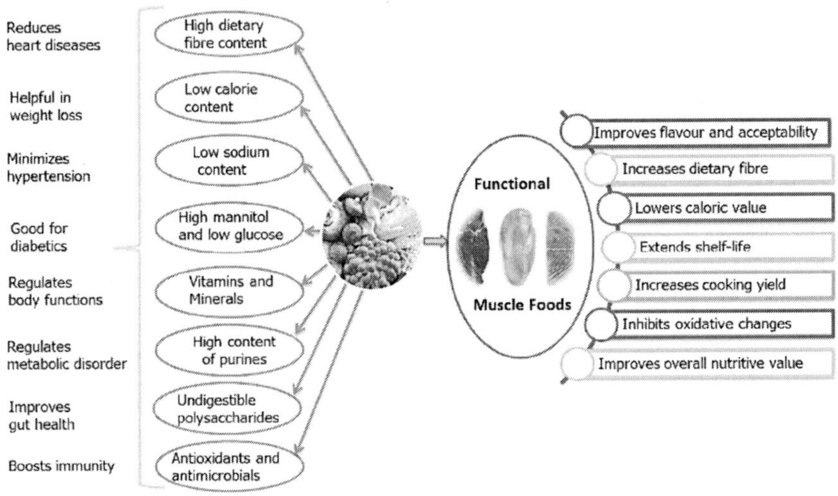

Figure 2. Schematic diagram showing the influence of mushroom nutrients on quality attributes of muscle foods vis-à-vis human health effects.

Over the years, a number of researchers have successfully incorporated mushrooms and its various parts (stipes and stem wastes) in formulation of various muscle food products like chicken sausages, salted cooked beef, tuna meat, kuruma shrimp, emulsion-type pork sausages, traditional Turkish meatball, fermented pork sausages, sutchi catfish pattiesetc. A summary on the effects of edible mushrooms on physico-chemical properties, colour and oxidative stability, shelf-life and sensory attributes of muscle food products is presented in Table 2.

Table 2. Effects of edible mushrooms and its parts on quality attribute of functional muscle food products.

Mushroom Variety and Level Used	Types of Muscle Food	Quality Parameters and Storage Conditions	Effects
Dried mushroom (*Pleurotus ostreatus*) @ 4%, 8% or 12%	Beef patties	Quality attributes (stored at −18 to −20 °C for 6 months)	• Increased protein, fat and ash contents, water holding capacity • Decreased moisture, carbohydrate contents, pH value, tenderness, plasticity, cooking loss • Better organoleptic properties of patties at 4 and 8% level
Mushroom (*Agaricus bisporus* or *P. ostreatus*) powder @ 5% or 10%	Traditional Turkish meatball	Sensory and physical (colour and texture) analysis	• Imparted positive effect on hardness • Meatball with *P. ostreatus* at 5% level was the best-liked
Mushroom (*Boletus edulis*) extract @ 1%, 3% or 5%	Beef burger	Antioxidant activities (stored at 4 °C for 8 days)	• Protected lipid peroxidation • Protected arachidonic (c20:4n6) and eicosapentaenoic (c20:5n3) acids • Extended shelf-life
Winter mushroom (*Flammulina velutipes*) powder @ 0.5, 1%, 1.5% or 2%	Emulsion-type pork sausages	Quality parameters (pH, lipid oxidation, texture and sensory properties)	• Increased water holding capacity and pH at >1% inclusion level • Decreased exudation of fat and water from the sausages • No adverse effect on colour and sensory properties at ≤1.5% inclusion level • Could replace phosphates in meat products • Had soft texture compared to control samples with phosphate
King oyster mushroom (*Pleurotus eryngii*) powder @10% or 15%	Chicken burger	Physical properties and sensory evaluation (stored at 6 ± 1 °C for 7 days)	• Increased water holding capacity • Lowered weight loss during cooking and thickness during storage at 15% inclusion level • Improved juiciness and tenderness
Winter mushroom (*F. velutipes*) powder @ 0.5% or 1%	Low-salt chicken sausages	Sensory analysis (stored at 4 °C for 3 days)	• Increased pH of the meat batter • Inhibited lipid oxidation, softened texture • Improved the nutritional quality • No negative effect on colour and sensory properties
Blanched grey oyster mushroom (*P. sajor caju*) @ 25% or 50%	Chicken patties	Optical and textural properties (colour, textural and cooking characteristics)	• Decreased lightness, yellowness but no change in redness values • Had similar cooking yield and moisture contents compared to the all-meat control • Less hardy, cohesive, chewy, and gummy
Mushroom (*F. velutipes*) extract @ 1 mL equivalent to 3 mg ergothioneine	Bigeye tuna (*Thunnus obesus*) meat	Antioxidative activity and anti-discolouration efficacy (frozen at −70 °C for 3 months)	• Showed higher radical scavenging activity • Suppressed lipid oxidation • Significantly inhibited the formation of metmyoglobin • Overcame the browning of big eye tuna meats up to 7 days of storage • Plays an important role as a colour stabilizer of meats
King oyster mushroom @ 20%, 30%, 40%, or 50%	Cuttlefish (*Sepia esculenta*) surimi gel	Physico-chemical and sensory attributes	• Significantly decreased the hardness, cohesiveness, and gumminess. • Increased the springiness of paste • Improved the nutritional properties and functionality of surimi gel

Mushroom (*A. bisporus*) powder @ 2%	Beef meat emulsion	Rheological and structural characteristics	• Exhibited improvement in textural properties • Improved viscoelastic behaviour • Exhibited higher heat resistance and emulsion strength • Provided a higher protein adsorption at the lipid interface, resulting in a well-ordered emulsion structure
Ground white mushroom (*A. bisporus*) @ 50% or 80%	Meat based dish (beef taco blend)	Nutritional quality and flavor profiles	• Improved nutritional quality by reducing calories, saturated fat, and cholesterol • Enhanced the flavor of the reduced salt version of the tacos • Substitution with mushrooms @ 50% or even 80% can maintain flavour intensity scores even with less salt like that of all-meat, full salt control
Mushroom (*A. bisporus*) @ 0.5, 1% or 2%	Sucuk (dry-fermented product using beef meat (90%) and beef fat (10%)	Quality properties during ripening period at (18 ± 2 °C) up to 12 days (stored at 4 ± 1 °C in refrigerator for 60 days)	• Prevented lipid oxidation • Did not affect the sensory attributes • Had a significant effect on the free fatty acids, peroxide value, appearance and colour indices • Improve the product quality
Pleurotus eryngii, F. velutipes, *Lentinula edodes*, *Pleurotus cornucopiae* and processing waste of *F. velutipes* @ 1% of each extract separately	Yellow tail (*Seriola quinqueradiata*) dark muscle	Colour stabilizing effects (storage in cold room at 5 °C and replaced daily with fresh ice)	• Had antioxidative activities against lipid oxidation and metmyoglobin formation. • Exerted anti-discolouration efficacy and maintained bright-red colour after 2 days of ice storage
Oyster mushroom (*P. sajor caju*) powder @ 2%, 4% or 6%	Chicken sausages/frankfurters	Nutritional composition and textural properties (stored at −18 °C until analysis)	• Enhanced dietary fibres up to 6.20% and ß-glucan up to 14.30% significantly • Lowered fat content but unchanged adhesiveness and cohesiveness attributes
Button mushroom (*A. bisporus*) @ 15%	Patties from sutchi catfish (*Pangasius hypophthalmus*)	Physical, chemical, microbial and sensory properties (stored at 6 ± 2 °C for 20 days)	• Increased the nutritional quality • Lowered the lipid hydrolysis development rate • Significantly lowered total volatile base nitrogen content • Significantly reduced total plate count • Improved shelf-life (up to 16 days) compared to control
Shiitake mushroom (*L. edodes*) powder @ 2%, 4% or 6%	Pork patties	Consumer acceptability and perception (stored at −20 °C up to one week)	• Declined moisture loss • Improved texture, juiciness, overall acceptability • Increased mushroom flavour and juiciness while still deemed just as acceptable as the all-meat patties

White jelly mushroom (*Tremella fuciformis*) @ 10%, 20% or 30%	Pork patties	Quality and sensory characteristics	• Better moisture and fat-holding capacity • Had significantly higher cooking yield • Had a higher lightness (64.31~67.23) and yellowness content than control • No adverse effect on appearance, colour, flavour and texture
Mushroom (*A. bisporus*) @ 15% or 30%	Beef burger	Physico-chemical and sensory measurements (vacuum packed and stored at −18 °C)	• Modified the texture, moisture and water activity • Had a significantly lower hardness than the control • Did not affect colour of the product
Mushroom (*F. velutipes*) extract @ 1% or 10%	Yellowtail (*Seriola quinqueradiata*)	Colour stability and lipid oxidation (stored at 0–2 °C for 4 days)	• Remarkably suppressed the browning development in dark muscle • Improved the quality of fish meat • Significantly increased shelf-life of dark muscle in terms of colour • Effectively controlled meat discolouration due to myoglobin and lipid oxidation
Button mushroom (*A. bisporus*) @ 5%, 10%, 15% or 20%	Fish paste	Texture profile analysis	• Increased the elasticity, hardness, brittleness, and gumminess • Decreased lightness and increased redness with increasing level • Highest overall acceptance at 10% supplementation
King oyster mushroom (*P. eryngii*) @ 5%, 10%, 15% or 20%	Silver white croaker (*Pennahiaargentata* fried fish cake	Quality properties and sensory characteristics	• Highest values in terms of strength, hardness, gumminess and brittleness • Decreased the degree of lightness • Increased redness and yellowness with increasing level • Exhibited better colour, flavour and overall acceptance at 10% level
Dried mushroom (*A. bisporus*) @ 1%, 2% or 4%	Beef patties	Shelf-life and changes in lipid and protein oxidation (stored at 4 °C for 16 days)	• Inhibited formation of lipid oxidation compounds • Lowered malondialdehyde and volatile aldehydes compared to control samples • Lowered the loss of thiols and tryptophan fluorescence @ 1% inclusion whereas at higher levels increased protein oxidation products compared to control • Extended shelf-life in a concentration dependent manner

Enoki mushroom (*F. velutipes*) stem waste powder @ 2%, 4% or 6%	Goat meat nuggets	Physico-chemical qualities, antioxidant capacity and lipid oxidation stability (stored at 4 °C up to 12 days)	• Improved cooking yield, mineral and dietary fibre content • Extended shelf-life of meat products by inhibiting lipid oxidation • No adverse effect on sensory attributes
Shiitake (*L. edodes*) by-products -stipes extract @ 0.3% or 0.6%	Fermented sausage (70% pork meat and 30% pork back-fat)	Quality characteristics, lipid oxidation and microbial stabilities (stored at 15 °C up to 30 days)	• Lowered ultimate pH values with higher number of lactic acid bacteria • Did not affect colour, texture and sensory quality • Improved lipid oxidation and microbial stabilities as well as controlled the growth of pathogens (*Staphylococcus aureus*, *Listeria monocytogenes*, and *Escherichia coli* O157 at 0.6%)
Shiitake (*L. edodes*) by-products- stipes extract (aqueous or ethanolic)	Fermented sausages (80% pork ham and 20% pork fat)	Antioxidant and microbial abilities (stored at 15 °C up to 40 days)	• Had higher antioxidant activities and inhibitory capacity against lipid oxidation • Had stronger antimicrobial activities against pathogens • Ethanolic extract partly showed beneficial effects on the flavour and taste improvements than control • Had significantly higher colour, flavour, taste and acceptability scores compared to control
Immature white button mushroom (*A. bisporus*) @ 10%, 20%, 30%, 40% or 50%	Ground beef (80/20 blend)	Physical and sensory characteristics (stored at −18 °C in between sheets of wax paper in plastic bags)	• No difference in yield, liking scores, lightness (L^* value), and red colour (a^* value) • Increased the moisture and yellow colour (b^* value) • Decreased the mechanical properties, sodium content, and fat content of the final products • Potential to reduce sodium contents in meat products providing a healthier product

Effect of Mushrooms on Quality Aspects of Muscle Food Products

Mushrooms on the Physicochemical Properties of Muscle Foods

The quality and acceptability of muscle foods depends on a number of different physicochemical properties including chemical composition, pH, water-holding capacity (WHC), emulsion stability, and cooking yield. The pH of muscle foods is particularly important because it influences their WHC, juiciness, cooking yield, texture, and shelf-life by regulating microbial growth. Studies have shown that incorporation of winter mushroom (F. velutipes) powder into emulsion-type pork sausages increased their pH, WHC, and cooking yield by decreasing the exudation of fat and water from the sausages . Similarly, Cha et al. reported that the incorporation of white jelly mushroom (T. fuciformis) into pork patties significantly increased their oil retention and cooking yield. Moreover, the introduction of 25% fresh mushroom into chicken patties has been reported to give an improved moisture retention (77%) and cooking yield (81%). The increase in cooking yield and water/fat retention are not only beneficial from technological and sensorial, but also from an economic viewpoint. In a recent study, the inclusion of enoki mushroom (F. velutipes) stem wastes into goat meat nuggets was shown to give a higher pH, emulsion stability, cooking yield, and WHC. Likewise, Bao et al. reported a slight increase in the pH of beef and fish products after addition of enoki mushroom extracts. The observed increase in pH after mushroom addition could be due to the relative abundance of basic amino acids compared to acidic amino acids in these products, as well as the natural buffering capacity of the mushroom proteins. In contrast, shiitake (L. edodes) extracts were found to reduce the pH of fermented sausages during 30 days storage at 15 -C , which may have been due to the presence of lactic acid bacteria that generated acids within the fermented sausages.

The addition of mushroom also influences the chemical composition and nutritional profil of muscle foods. As discussed earlier, these effects can be attributed to the presence of relatively high levels of protein, minerals, and dietary fibres in mushrooms. The incorporation of dried mushroom (*P. ostreatus*) into beef patties has been reported to increase the protein, fat, and ash content of the end product . Similarly, the incorporation of dried grey oyster mushrooms significantly decreased the fat content of cooked chicken patties . The incorporation of dried enoki mushroom extracts into goat meat nuggets has been reported to increase their dietary fibre and ash contents . Wan Rosli et al. reported that the incorporation of oyster mushroom (P. sajor caju, PSC) powder into chicken meat decreased the fat content but increased the dietary fibre content in frankfurters in a dose-dependent manner.

Taken together, these studies clearly show that the composition of muscle-based foods can be manipulated by adding different types and amounts of mushrooms during processing. Incorporation of mushrooms either in the form of powder or extract influences the physicochemical properties, sensory attributes, and nutritional profiles of muscle foods.

Mushrooms on Lipid Oxidation in Muscle Foods

The oxidation of the lipids and proteins in muscle foods is undesirable because it leads to rancid odours, off-flavours, discolouration of the products. Again, lipid oxidation by different means produces free radicals (such as alkyl, alcoxyl, and peroxyl radicals) which have been observed to induce protein oxidation. For example, cooking or heating muscle food products above 60 °C initiate oxidative cleavage of the porphyrin ring, resulting in release of heme iron, which can lead to increased lipid and protein oxidation. These oxidative reactions are extremely complex and lead to the loss of valuable nutrients, as well as the generation of numerous types of reaction products . For instance, the essential amino acids and fatty acids may be lost, whereas volatile off-flavours are produced.

Many factors influence the lipid and protein oxidation, including oxygen, temperature, light, and transition metal ions. Researchers are increasingly trying to identify and utilize natural antioxidants to inhibit lipid and protein oxidation in muscle foods . The utilization of these antioxidants helps in improving the food quality, shelf life, and nutritional profile.

Mushrooms contain a wide range of natural antioxidants, including phenolic compounds, ergothioneine, ascorbic acid, tocopherols, and carotenoids. The fruit bodies and mycelium of mushrooms also contain various types of natural antioxidants, including glycosides, polysaccharides, selenium, ascorbic acid, tocopherols, and carotenoids. Researchers have reported that phenolic compounds (3–11 mg/g) and flavonoids (2.5–4.8 mg/g) are the major bioactive compounds responsible for the antioxidative activity of the fruit bodies of edible mushrooms. Winter mushrooms (*F. velutipes*) are also known to have strong antioxidant activity because they contain phenolic compounds such as quercetin, chlorogenic acid, gallic acid, proto-catechuic acid, and flavonoids. Methanol extracts of B. edulis have also been reported to contain constituents that are known antioxidants, such as ascorbic acid (18.7 mg/g dw), tocopherols (18.7 mg/g dw) and phenolic acids (9.74 mg/kg dw) . Waste materials (stipes) from shiitake (L. edodes) mushroom have also been reported to contain various kinds of natural antioxidants . Acetone and methanol extracts of various mushrooms, including Amanita rubescens, Lepista nuda, Cantharellus cibarius, Hypsizigus marmoreus, Lactarius piperatus, Polyporus squamosus, Mucor circinelloides, Russula cyanoxantha A. bisporus, L. edodes, and V. volvacea have also been reported to exhibit strong antioxidant activity, which is mainly attributed to their high levels of phenolics and flavonoids.

Mushroom extracts from the fruiting bodies of edible mushrooms (*F. velutipes*) have been reported to inhibit lipid and protein oxidation in raw beef and fish (bigeye tuna) during storage, which was attributed to the presence of natural antioxidants . Alnoumani et al.incorporated dried A. bisporus powder into salted cooked ground beef and tested its ability to protect the lipids and proteins from oxidation during storage.

The mushroom extract was found to effectively inhibit lipid and protein oxidation, with about 88–94% lower malonaldehyde values and 99% lower volatile aldehydes being produced after 16 days storage compared to the control sample. In another study, Nayak et al. reported significantly lower oxidative changes in sutchi catfish (Pangasius hypophthalmus) patties after a button mushroom (*A. bisporus*) extract was incorporated, leading to an appreciable increase in shelf life. The antioxidative activity of mushrooms and their extracts have also been demonstrated in various other studies, including B. edulis extracts in beef burger patties, ground white mushroom in dry-fermented beef products , F. velutipes extract in bigeye tuna, F. velutipes extract in kuruma shrimp , and L. edodes extracts in fermented pork sausages.

These studies clearly show that mushrooms and their extracts contain a diverse range of natural antioxidants that can improve the quality and shelf-life of muscle food products by inhibiting lipid and protein oxidation.

Mushrooms on the Textural Properties of Muscle Foods

The quality attributes and acceptability of muscle foods is strongly influenced by their textural properties. Muscle foods are compositionally and structurally complex soft materials with semi-solid textures that influence their preparation, mastication, and digestion. The textural attributes of muscle foods is not only largely governed by the gel-forming and emulsification properties of the proteins they contain, but also effect by other components, such as lipids and minerals. The incorporation of mushrooms into muscle foods influence their rheological characteristics, which must be accounted for when designing mushroom-enriched products. One of the advantages of using mushrooms in meat products is that they already have quite meat-like textures themselves because they have a very firm texture, and their dietary fibre fractions form a dense meaty texture when processed with muscle foods. As a result, they can often be incorporated up to a certain percentage into muscle foods without causing major adverse effects on their textural attributes . Changes in quality aspects especially textural attributes of muscle foods due to incorporation of mushroom can be assessed from different studies is listed in Table 2.

A number of researchers have examined the impact of mushrooms on the textural properties of muscle foods. Choe et al. reported a decrease in hardness, springiness, gumminess, and chewiness of sausages after incorporation of a mushroom powder.

Similarly, Banerjee et al. reported that the hardness, springiness, cohesiveness, and gumminess of meat nuggets decreased after addition of mushroom extracts, but these effects were not statistically significant. The hardness and other textural attributes of chicken patties have been reported to decrease after replacement of 25% or 50% of chicken meat with oyster mushroom . Decreased hardness, cohesiveness, and gumminess but increased springiness have been reported when king oyster mushroom is incorporated into surimi gel prepared from cuttlefish (*Sepia esculenta*) meat paste .

In general, these results suggest that incorporation of mushrooms into muscle foods usually leads to softening of the final products. There are a number of possible physicochemical phenomena that may account for this effect. For instance, mushrooms contain relatively high levels of dietary fibres that can form a 3D biopolymer network that traps fluids, thereby leading to a softer texture in the muscle food products . Moreover, the incorporation of high levels of mushrooms into muscle foods reduces the concentration of solubilized muscle proteins, thereby decreasing their ability to form strong gels.

Mushrooms on the Appearance of Meat Products

The appearance of muscle foods, such as their opacity, colour, and surface sheen, provide a visible indication of their quality and freshness, thereby playing an important role in determining consumer purchasing decisions. The incorporation of mushrooms into muscle foods may impact their appearance in a number of ways. Mushrooms naturally have a different colour to meat or fish and, therefore, blends will have a different appearance than meat or fish alone. Moreover, the mushroom extracts may contain particles that have sizes and shapes that are different from those present in muscle foods, which can alter their visual texture. Finally, mushrooms contain antioxidants and other molecules that may inhibit colour changes in muscle foods. In this section, we provide a brief overview of studies that have examined the impact of mushrooms on the appearance of muscle foods.

Bao et al.monitored changes in the met-myoglobin concentration and colour of minced beef and bigeye tuna meat products containing winter mushroom extract during cold storage. The authors reported that the presence of the mushroom extracts significantly decreased the met-myoglobin concentration in the meat products. As a result, the desirable colour of the beef and tuna products was maintained for up to 12 and 7 days storage compared to 6 and 2 days for the control samples (no mushroom), respectively. The colour stabilizing effect of the mushroom extract was attributed to the presence of ergothioneine, which reduced the rate of met-myoglobin formation, which is known to promote discolouration in muscle foods. Similarly, an ergothioneine extract from mushroom was shown to prolong the stability of the red colour of yellowtail and tuna fish meat during cold storage .

In another study, researchers reported that the presence of 50% oyster mushroom in cooked chicken patties did not affect their redness (a*) but did decrease their lightness (L*) and yellowness (b*). Incorporation of white winter mushroom powder into emulsion-type pork sausages was reported to have little impact on their appearance.

Similarly, incorporation of up to 20% A. bisporus mushroom into beef burgers did not strongly affect their appearance. The introduction of shiitake mushroom powder into frankfurters had little impact on their initial colour but did increase their yellowness during storage. In a study on cooked beef taco, Wong et al.reported a decrease in lightness (L*) with increasing level of mushroom (25–75%). In addition, the incorporation of 75% mushroom into the taco meat led to a lower redness (a*) than the all-meat control samples. These effects might arise due to the fact that the mushrooms were darker than the meat products, and because the myoglobin content in the final products was reduced after addition of the mushrooms.

In contrast to the above studies, the addition of white jelly mushroom was found to slightly decrease the redness and increase the yellowness of cooked pork patties, which might be due to the transparent white colour of these mushrooms. In general, the impact of mushrooms on muscle foods depends on the initial colour of the mushrooms and muscle foods, as well as any physical interactions or chemical reactions that can occur between them.

Mushrooms on the Microbiological Quality of Muscle Foods

Muscle foods contain high levels of macronutrients and micronutrients that spoilage or pathogenic microorganisms can utilize to grow. Therefore, it is important to have effective strategies to extend the shelf-life and ensure the safety of this type of food . Many mushrooms and their constituents exhibit antibacterial and antifungal properties.

Consequently, their incorporation into muscle food products may have the added advantage of improving their safety and shelf-life. These antimicrobial properties have been attributed to a number of different mushroom constituents, including high-molecular weight (peptides and proteins) and low-molecular weight (terpenes, steroids, anthraquinones, benzoic acid derivatives and quinolones) compounds secreted by the mushroom fruiting body for its own survival . Medicinal mushrooms like Aleurodiscus, Coprinus, Clitocybe, Daedalea, Marasmius, Merulius, Pleurotus, Polyporus, Poria, Psathyrella, and Tricholoma spp. have been used as a source of natural antibiotics for the treatment of various types of disease because of the immunomodulatory properties of _-glucans and the antibacterial properties of secondary metabolites . In this section, we review studies on the antimicrobial properties of mushrooms and their constituents, with a focus on their application in muscle foods.

Chowdhury et al.reported that extracts of L. edodes, P. ostreatus and Hypsizigus tessulatus demonstrated antimicrobial activity against all bacteria and fungi tested, with minimum inhibitory concentration (MIC) values ranging from 1 to 9 mg/mL. In this study, the L. edodes extracts were more effective than those from the other two mushroom species. Similarly, shiitake (L. edodes) mushroom extracts, isolated using organic solvents and supercritical fluids, exhibited antibacterial effects against pathogenic microorganisms such as Streptococcus pyogenes and Staphylococcus aureus , whereas extracts isolated using only supercritical fluid revealed antimicrobial activity against Micrococcus luteus and Bacillus cereus . Extracts from Pleurotus florida mushroom have also been reported to exhibit strong inhibitory effects on the growth of both Gram-positive and Gram-negative bacteria and could, therefore, be considered as an alternative to traditional antibiotics.

Other studies suggest that mushrooms or their extracts can be used as natural preservatives that improve the shelf-life of foods by inhibiting the growth of spoilage microorganisms in muscle foods. Button mushrooms have been reported to exhibit antimicrobial activity in sutchi catfish patties, leading to a considerable extension in their shelf-life .

In another study, shitake (L. edodes) extracts were shown to exhibit good antimicrobial activity in fermented sausages, thereby extending their shelf-life and inhibiting the growth of pathogens such as S. aureus, Listeria monocytogenes, and Escherichia coli O157. Stojkovi ´c et al.reported that a methanolic extract from Boletus aereus was effective at controlling microbial growth against food-poisoning organisms in pork meat, including S. aureus, L. monocytogenes, *E. coli*, and Salmonella Typhimurium. Overall, these studies show that incorporating mushrooms into muscle foods, like meat or fish, may be an effective means of increasing their shelf-life and safety.

Mushroom on Sensory Attributes of Muscle Foods

The sensory attributes of food products play an important role in determining their quality and desirability. The overall sensory impression of muscle foods depends on their appearance, flavour, texture, and oral processing. The incorporation of edible mushrooms into meat and fish products changes these physicochemical characteristics to an extent that depends on the type and level of mushrooms used, thereby altering their sensory attributes. Choe et al. reported that incorporation of 1% winter mushroom powder into emulsion-type pork sausages led to better sensory scores (texture, flavour, and acceptability) than sausages with 2%. Wan Rosli et al. reported that incorporation of 25% to 50% oyster mushrooms into chicken patties led to blended products that had similar sensory and liking scores as the all-meat control patties. Likewise, incorporation of 25% oyster mushrooms into meat patties was also shown to have no adverse effects on their sensory attributes.

Myrdal Miller et al. found that incorporating 80% of white button mushrooms into reduced-salt ground beef tacos did not impact their flavour intensity. This study suggests that in some cases it is possible to include high levels of mushrooms into muscle food products while maintaining their desirable flavour profiles. This effect may be due to the relatively high content of free amino acids in mushrooms, which generate desirable meat-like umami, sweet, and bitter tastes. The umami taste makes the edible mushrooms palatable and adaptable in most food preparations . Chun et al. reported that incorporating different levels (2%, 4% and 6%) of shiitake mushroom powder into pork patties increased their firmness, juiciness, flavour, and overall acceptability in a dose dependent manner. Similarly, incorporation of king oyster mushroom (30%, 40%, and 50%) into cuttlefish (*S. esculent*a) paste led to higher overall acceptability scores than the control paste. A number of other studies have also reported no changes or an improvement in the sensory attributes of muscle food products after addition of mushrooms or their extracts, e.g., enoki mushroom powder (2%, 4% and 6%) to mutton nuggetsand ground white jelly mushrooms (10%, 20%, and 30% ground) to pork patties .

Conclusions

Mushrooms are a rich source of important nutrients and bioactive components, including proteins, fibres, vitamins, minerals, and nutraceuticals, while also being low in calories, sodium, fat, and cholesterol. They can also be produced more sustainably than meat products, with less damaging effects on the environment. Moreover, they have many flavor and textural attributes that are compatible with meat and fish products.

Consequently, they are an extremely valuable functional ingredient for creating muscle food products with enhanced nutritional and sustainability profiles. In particular, they can be used to create foods designed for a flexitarian diet where a fraction of the meat or fish in a product is reduced.

There are many species of edible mushrooms that have not yet been explored for their potential application in food products. These novel mushroom species may have different nutritional, sensory, and environmental impacts than those exhibited by the species used so far. Consequently, there is a need for research on these other species.

Moreover, further research is required to identify the mechanisms of action of different constituents in mushrooms, e.g., their antimicrobial activity, antioxidant activity, structures, textural attributes, and flavour profiles. Given the increasing demand of consumers for healthier and more sustainable foods, it is likely that the utilization of mushrooms within muscle foods is likely to continue to grow.

Chapter 24

Value Addition of Mushroom

Mushrooms have huge health and nutritional benefits and can solve many problems of under-nutrition and malnutrition. Despite this fact mushroom cultivation and its utilization is not catching up fast because mushrooms are highly perishable. Thus it is important to process mushrooms into value added products which will not only cater to the protein and micronutrient requirement of masses but at the same time will solve the problem of short shelf-life and postharvest losses of mushrooms. Following are some technologies to process mushrooms into value added products with extended shelf life. Many persons do not purchase mushrooms as they don't know how to cook these. Hence we will read about some recipes.

Generally, for long-term storage of mushrooms, canning and drying processes along with some value addition technology are employed. Among the various techniques employed for preservation of mushrooms, drying seems to be an effective approach to extend shelf life and ensure distribution. Various drying techniques are employed to dry different food products. Each technique has its own advantages and limitations . Studies comparing traditional sun drying and other drying techniques show that the use of solar dryer leads to a considerable decline of the drying time and to a significant development of the product quality in terms of color, texture and taste. Canning is the most common process used for mushrooms. Preservation of mushrooms by canning has turned out to be considerably more specialized in recent years .

Dehydrated mushrooms are used as an important ingredient in several food formulations including instant soups, pasta, snack seasonings, casseroles, and meat and rice dishes. Mehta et al,developed various products of *Agaricus bisporus* like mushroom candy, mushroom ketchup, murraba, mushroom chips, mushroom nuggets, and mushroom soup- powder, mushroom biscuits, ready- to- eat mushroom curry. Sharma et al.also successfully prepared biscuits from mushrooms. Ulziijargal et. alstudied the quality of bread supplemented with mushroom mycelia and found that the substitution of 5% wheat flour with mycelium powder did not adversely affect the texture profile of the bread. Also, after baking, mycelium-supplemented bread was found to have substantial amounts of GABA and ergothioneine. Desayideveloped mushroom fortified noodles and found that the amongst the different treatments 10 per cent mushroom powder along with 0.2 percent vanilla flavour recorded highest scores for organoleptic parameters like colour and appearance, flavour, crispness, taste and overall acceptability even upto 30 days of storage. These studies indicate that medicinal mushrooms have much to offer to healthcare of common man as well as armed personnel. The value added products can be prepared from fresh or dried mushrooms. Method of preparation of various products are described below.

Mushroom Flour: The flow diagram of standard procedure is as follows:

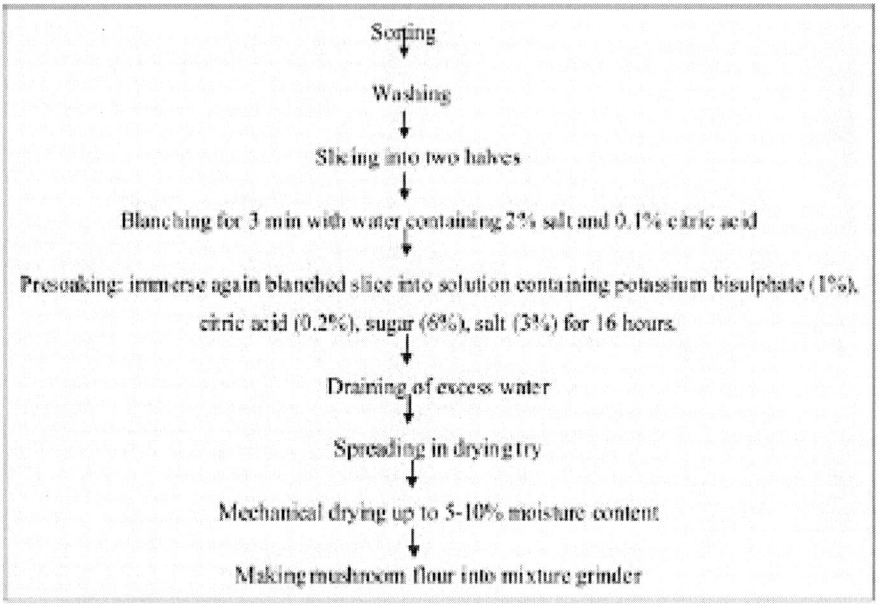

Fig. 1. Flow diagram showing preparation of mushroom flour

Preservation of Mushrooms

Drying and canning are the two most popular methods of preservation. For short time preservation, we may also keep mushrooms in salt solution (steep preservation).

Drying of Mushrooms

As you are aware that drying is one of the most important methods to decrease the water content of the produce. The dried produce can be further utilized for preparing a number of value added products. Button mushroom cannot be easily sundried. However, mushrooms like oyster, shiitake, wood ear mushroom, etc can be easily sundried (Fig. .1). Besides sun drying, mushrooms can be dried in cabinet dryers at a drying temperature of 55-60°C which gives dehydrated final product of lower moisture content with longer shelf life and better quality.

Fig. .1: Dried oyster mushroom

Mushrooms have about 90% water and hence after drying these become very light.

Care has to be taken while sun-drying. Mushrooms should be dried in dust free areaand in containers where these don't get dispersed with wind. Additional advantage of sun drying is the increase in Vitamin D content in the mushroom. Mushrooms having open gills should be firstly dried with gill side upwards preferably in shade and then dried in sun.

You may use solar based heater. But it is important to monitor the temperature of drying. Temperature above 60°C is not desirable as it leads to charring of sugars, loss of flavour and also decrease in rehydrability (that is gain of weight on rehydration of the dry product).

You may powder the dried mushrooms and use it to make various products. Up to 10% powder can be added to flour that may be used to make biscuits, bread, cake or any other bakery product.

Canning of Mushrooms

Canning is a technique by which the mushrooms can be stored for longer periods up to a year and most of the international trade in mushrooms is done in this form. Mushrooms can be canned whole, sliced and stems and pieces as per demand. The canning process involves cleaning, blanching (5-6 minutes at 95-100°C), filling into can with brine solution (2% salt with 0.1% citric acid or 100 ppm ascorbic acid), sterilization by heat (118°C), cooling, labelling and packaging (Fig. .2).

Fig. .2: Canning of mushroom

Steeping Preservation of Mushrooms

This method is simple and economical and the mushrooms can be preserved for short period by steeping them in solution of salt or acids. Solution consisting of 2%,sodium chloride, 2% citric acid, 2% sodium bicarbonate and 0.15% KMS is used for steeping preservation of blanched mushrooms for 8-10 days at 21-28°C (Fig. .3).

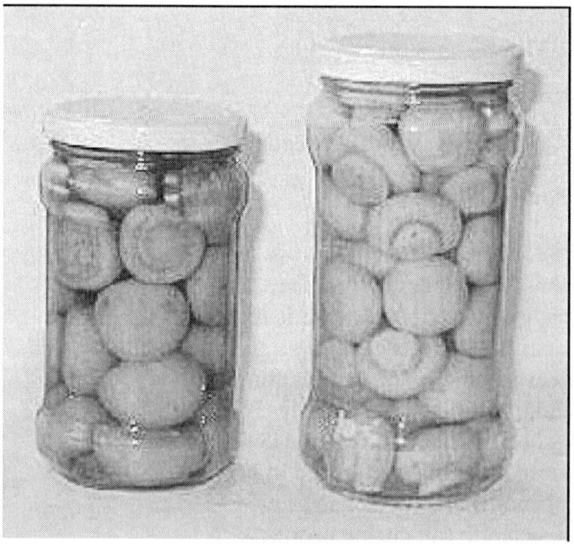

Fig. .3: Steep preservation

Value Added Products of Mushrooms

You can also prepare various value added products such as mushrooms pickle, jam, sauce, candy, preserve, chips etc. from fresh mushrooms whereas from the dried mushroom powder value added products like instant soup mix, bakery products, papad, nuggets etc. can be prepared (Fig. .4).

Mushroom Biscuit

Mushroom soup powder

Mushroom nuggets

Mushroom papad

Fig. .4: Different products made using mushroom powder

Conventional Products

Mushroom Pickle

Under our conditions, especially in rural area, it may not be possible to do canning of mushrooms. A locally accepted alternative is to make pickle. In fact, some mushrooms like milky mushroom are more accepted as pickle than as fresh product. Pickling of mushrooms is an easy home scale process for preservation of mushrooms to a value added product of high market acceptability. For preparing mushroom pickle, one kg mushrooms are washed in 0.03 to 0.05% KMS solution and then blanched for 5 min at 85°C. The blanched mushrooms are washed in cold water 2-3 times and the excess water is drained off. Mushrooms are cut in halves or quarters according to the size. Then the mushrooms are subjected to salt curing process, in which 20 g sodium chloride per kg of mushroom is added and kept overnight. The excess water that oozes-out of mushroom is removed on the next day. Mushrooms are allowed to dry for 2-3 hours and spices, salt and preservatives are mixed to the desired taste and quality of mushroom pickle. We add heated mustard oil from the top to the mixture and blend with both hands, leave overnight for imbibing of spices and flavour development, fill in plastic/glass jars and again top-fill with mustard oil, close the caps and seal the jars and store pickle at shaded ambient place.

For 1 kg mushroom you need various spices *viz.*, turmeric powder (20g), black mustard seed powder (35g), red chilli powder (10g), cumin seed powder (1.5 g), carom seed (10 g), nigella seed (kalonji) (10 g), fennel seed powder (1.5 g), salt (90 g) and mustard oil (200 ml). You may also use Acetic acid (upto 100 ml) and sodium benzoate within the permitted limits as preservatives. This pickle can be stored up to one year in the airtight bottles.

Given below is another recipe for making sweet-sour pickle of mushroom:

Sweet-sour/mixed mushroom pickle

Ingredients

Mushrooms 250 g

Onion 50 g

Ginger 25 g

Garlic 2-3 cloves

Mustard oil 100 g

Salt 1 spoon

Red chilies 1/2 spoon

Garam masala 1/2 spoon

Jaggery 100 g

Vinegar /acetic acid 3 cup/ 1 spoon

Mustard seeds 10 g

Method

1. Wash the mushroom and cut into small pieces.
2. Cut ginger, onion garlic into fine pieces.
3. Grind the mustard seeds coarsely.
4. Make a mixture of Jaggery and vinegar.
5. Heat the mustard oil and fry the mushrooms, keep fried mushrooms to one side.
6. Fry onion ginger, garlic till light brown.
7. To this add fried mushrooms and solution of jaggery and vinegar.
8. Then add mustard powder, chillies, salt and garam masala. Mix well and keep in sun for 2-3 days.

Mushroom Preserve (Murabba)

For preparing mushroom preserve, fresh button mushrooms are graded, washed, pricked and blanched in 0.05% KMS solution for 10 min. Blanched mushroom are then dipped in 50 °Brix sugar solution and refrigerated overnight. Next day mushroom are strained out of sugar solution and to the solution we add 0.1% citric acid and sufficient sugar and heat to attain strength of 60 °Brix. Mushrooms are then again dipped into it and kept overnight. This process is repeated to raise the concentration of syrup to 70 °Brix and mushrooms are dipped into it for 1 week to prepare preserve. The preserve is then drained out of sugar syrup and filled in a container with freshly prepared sugar syrup of 68 °Brix. The containers are then sealed airtight and stored.

Mushroom Candy

A fruit or vegetable impregnated and coated with sugar, subsequently taken out and dried is called a candied fruit or vegetable.

- ❖ Wash fresh mushrooms after harvesting and halved longitudinally into two pieces.
- ❖ Blanch halves of mushrooms for 5 min in 0.05% of KMS solution and treat with sugar for half an hour after draining.
- ❖ Give sugar treatment at the rate of 1.5 kg sugar per kg of blanched mushroom.
- ❖ Initially sugar has to be divided into three equal parts. On the first day, cover blanched mushrooms with one part of sugar and kept for 24 h. Next day, cover the same mushrooms with second part of sugar and are kept overnight and on the third day remove mushrooms from the sugar syrup (boil 3rd part of sugar and 0.1% of citric acid to bring its concentration up to 70°Brix).
- ❖ Mix the mushrooms with this syrup and again boil the contents for 5 min to bring its concentration up to 72°Brix.
- ❖ After cooling, remove the mushrooms from the syrup and drained for half an hour.
- ❖ Place the drained mushrooms on the sorting tables to separate, to reject defective and unwanted pieces.
- ❖ Finally subject the mushroom pieces to drying in a cabinet dryer at 60°C for about 10 h. As soon as these become crispy, take out all mushrooms and packed in polypropylene bags and sealed.
- ❖ The mushroom candy can be stored up to 8 months with excellent acceptability and good chewable taste.

Mushroom Chips

For preparing mushroom chips, freshly harvested button mushrooms are washed, sliced (2 mm) and blanched in 2% brine solution. The mushrooms are dipped overnight in a solution of 0.1% of citric acid + 1.5% of NaCl + 0.3% of red chilli powder. After draining off the solution, the mushrooms are subjected to drying in cabinet dryer at 60°C for 8 hr. Then it is fried in the refined oil and good quality chips are prepared. After mixing the spices, the chips are packed in polypropylene packets and sealed.

Mushroom Biscuit

Both button or oyster mushroom can be used to prepare delicious and nutritious mushroom biscuits using ingredients *viz.*, refined wheat flour (*maida*) & mushroom powder (in 80:20 or 90:10 ratio), sugar (30%), ghee (bakery fats) (45%), baking powder (0.6%), ammonium bicarbonate (0.3%), salt (0.6%), milk powder (1.5%) and vanilla essence (0.02%). For making biscuits all the dry ingredients are finely ground and sieved. Then fat and sugar are mixed well for 5-7 minutes using Dough kneeder. These ingredients are then added to dough kneeder with other dry ingredients for dry mixing. Thereafter, water is added to make dough cohesive and homogenous and mixing is continued till fully done. Then dough is kept for 10 minutes covered with wet cloth. Thin sheets of dough (1.25 cm thick) are made and cut into different shapes of biscuits using different steel dies. These raw cut biscuits are then baked in hot oven (at 180°C) for 20 minutes and after cooling biscuits are ready for packaging.

Mushroom Soup Powder

For preparing mushroom soup powder, dried oyster or button mushrooms are finely ground in a pulveriser to pass through 0.5 mm sieve. This mushroom powder (20g) is then mixed with milk powder (25g), corn flour (40g), salt (8g), sugar (3g), black pepper (2g) and oregano (2g). This soup mix has to be mixed with 6 times quantity of water for the preparation of good quality mushroom soup with characteristic aroma and taste.

Mushroom Nuggets

For preparation of mushroom nuggets, mushroom powder (dried and coarsely ground mushrooms) is mixed with the Black gram (*Urad*) dal powder (1:8) and a paste is prepared by adding required quantity of water. Salt (2%) and red chilli powder (1%) are added to the prepared paste and round balls of 2-4 cm diameter are made. The prepared balls are spread over a steel tray and are sun dried. These mushroom nuggets can be straightaway deep fried and used as snacks or can be used in vegetable curry preparation.

Mushroom Papad

Papad is a thin, crisp disc-shaped Indian snack food usually made from seasoned batter of peeled black gram flour (urd flour), lentils, chickpeas, rice, tapioca or potato, fried or cooked with dry heat. Papads can be supplemented for protein with mushroom either in the form of paste or dried powder in the batter prepared from other sources as mentioned above. This can make papad a wholesome food with high protein content. The steps can be summarized as below:

- Wash harvested mushrooms in water (containing 300 ppm KMS).
- Blanch mushrooms in hot water at 85°C for 5 min and then dip immediately in cold water.
- Cut mushrooms into halves or quarters and make a paste in commercial/ kitchen mixer grinder.
- Boil potatoes in a pressure cooker or vessel, peel and crush/grate boiled potatoes to a paste consistency
- Blend potato paste and mushroom paste in 50:50 ratio and add spices (Salt, chilli powder/ black pepper, cumin etc as per taste).
- Add KMS powder to the blend @ 300 ppm.
- Spread the papad batter on a polyethylene sheet in drying trays in appropriate round size (10 cm diameter) with uniform thickness (1-2 mm).
- Keep the trays in sun for sun drying or in tray drier at 50°C for 5-6 hr to reduce the moisture content to 5%.
- Peel off the dried papads from polyethylene film and pack in sets on 10-20 papads in one poly-bag.
- Store in a cool and dry place.

Note: To consume the product, it has to be fried at 170°C for 60 to 90 sec in cooking oil.

Mushroom Fritter

Ingredients:

Mushroom - 500gm	Turmeric powder -1tsp	Jira powder - 1tsp
Besan - 250 gm	Oil - 300ml	
Chilli powder - 1tsp	Salt - to taste	

Method of preparation:

- Wash and cut the mushroom in to desired size and shape.
- Make batter by mixing besan with water, salt and spices.
- Dip the mushroom in to the batter and keep for sometime.
- Heat oil in a karahi and fry the mushroom pieces till golden brown.
- Remove from fire when fritter turn crisp golden brown.

❖ Serve hot with tomato sauce or chilli sauce.

Mushroom Paknam: (Baked Mushroom)

Ingredients:

Mushroom - 1Kg	Green chilli slice - 100 gm	Gram flour - 500 gm
Fermented dry fish - 100 gm (optional)	Nakupi leaf - 200 gm	Salt - 15 gm

Method:

- ❖ Wash and Shred mushroom in to small pieces.
- ❖ Wash green chilli and cut into pieces.
- ❖ Wash nakupi leaf and cut into pieces (4-5cm length).
- ❖ Grind coarsely slice green chilli with properly roasted fermented fish and salt.
- ❖ Mix with mushroom and nakupi leaf very well.
- ❖ And add gram flour to the mixture by addition of little water till the mixture blended very well.
- ❖ Put the mixture in turmeric leaf and wrap over. Again cover / wrap with banana leaf.
- ❖ Pressure-cook the mixture for 15 to 20 mins.
- ❖ Take out from pressure cooker and roast in pan / tawa by upside down in low flame till it cooks properly (another 15 to 20 mins).
- ❖ Remove banana leaf and also turmeric leaf.

Pancake

- ❖ Replace the wheat flour at 10%, 20% or 30% level for soy flour and at 10% for preparation of pancakes.
- ❖ Sieve the wheat flour, soy flour, and mushroom flour with sugar into a mixing bowl and pour milk in it.
- ❖ Mix flours and sugar with milk and add some butter (2 table spoon) in this mixture.
- ❖ Spread this mixture over the pan and cook it with the help of butter.
- ❖ Cook till it become golden brown on both the sides.
- ❖ Serve it with sugar syrup, lemon, orange poured on it.

Laddo

- ❖ Replace the wheat flour at 10%, 20% or 30% level for soy flour and at 10% for mushroom flour respectively for preparation of ladoo.
- ❖ Melt the ghee in a pan over the flame.
- ❖ Add mixture of wheat flour, soy flour, and mushroom flour in the pan and roast it by stirring constantly.
- ❖ Add jaggery, ground cardamom and groundnut into it. Mix it properly.
- ❖ Grease your hands with ghee and shape the mixture into laddos.

Pharra

- ❖ Replace the wheat flour at 10%, 20% or 30% level for soy flour and at 10% for mushroom flour respectively for preparation of pharra.
- ❖ Add water and salt in the flour (wheat flour, soy flour, and mushroom flour) and make dough out of it.
- ❖ Stuff Bengal gram flour in the small dough
- ❖ Boil stuffed dough in water till tender.
- ❖ Take out and cut into small pieces.
- ❖ Shallow fry it with chopped Onion, Curry leaves, Green chilies and Rai.
- ❖ Serve hot.

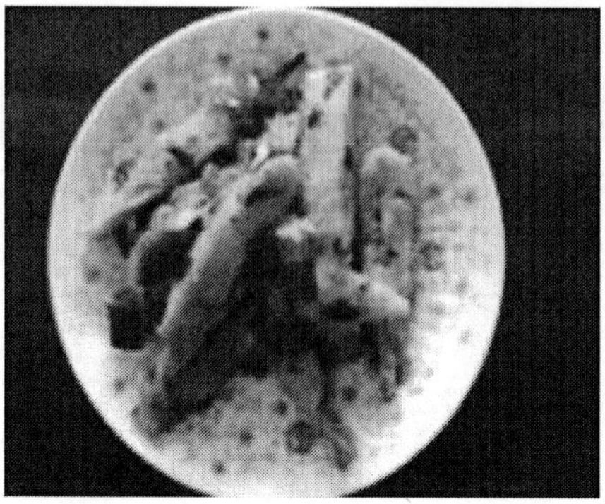

Mushroom Jam

- Make washed mushrooms pulp with mortar and pestle; and then sieve crude mixture pulp through an 80 nylon mesh.
- Process pulps from apple and banana into fruit jams according to the FAO guideline (FAO, 1997).
- Extract guava pectin by boiling the guava pieces (one part with two parts of water) for 30 min.
- Take Mushroom pulp 33-45 parts along with mixture of fruits such as apple and banana in 50% of the total mushroom for the preparation of jam.
- Determine Total Soluble Solids (TSS) and Total Titratable Acidity (TTA) in the pulp before jam preparation to establish any additional requirement to meet the recommended acid to sugar ratio.
- Add citric acid and sodium chloride to the mixture after analyzing sugar ratio
- Process the mixture to the desired TSS of 69% and check by using a hand Refractometer (Hand Refractometer 58 to 90° Brix).
- Finally, mix the jam was with class II preservative sodium benzoate (100 ppm). Later, fill the hot jam into sterilized glass bottles and left to cool and aluminum foiled. Then store it at room temperature (27 to 32°C).

Mushroom Squash

- Wash the mushroom with clean running water to remove external dust particles and to reduce the microbial load on the surface of the mushroom.
- Crush the mushroom in screw type juice extractor machine for extraction of juice.
- Add Amla (Indian gooseberry) juice and lemon juice along with the extract and filter the juices through muslin cloth.
- Dissolve a calculated quantity of potassium meta-bi-sulphite (750 ppm) in a part of water and mix well in the blended juice with the help of stirrer.
- Add sugar syrup made with the addition of citric acid, and water to the extracts at different proportion.

- ❖ Dilute the squashes with purified RO water until it reached 52° brix in Refractometer and adjust an acidity level of 0.16.
- ❖ Store the bottles containing squash at room temperature (28 ± 8°C) .

Novel Value Added Products of Mushrooms

Mushroom fortified corn extrudates: Fortification in extrudates to a level of 20% paste and 10% mushroom powder for both single and twin screw extruders can give extrudates of acceptable quality.

Mushroom fortified cakes: Mushroom fortified cake can be made by fortification to a level of 10-20% of dry mushroom powder in the wheat floor. The process of making cake remains the same.

Mushroom Based Vegetarian Sausages: Vegetarian sausages can be prepared from fresh mushroom by adding 5% saturated fat and binding agents such as carrageenan, soya protein concentrate, casein or xanthan gum.

Mushroom fortified instant noodles: Mushroom fortified Ready to cook instant noodles of acceptable quality and high nutritive value can be prepared by fortifying noodle dough with 4% of mushroom powder (*Pleurotus ostreatus*).

Mushroom corn extrudates

Mushroom fortified cake

Mushroom Based Vegetarian Sausages

Mushroom fortified instant noodles

Fig.10.5: Various novel products of mushroom

Mushroom Recipes

Many a times people do not purchase mushrooms as they don't know what to do with it, that is, how to cook it. Cooking mushrooms is similar to cooking of any other vegetable with slight difference that other vegetables on over cooking become soft whereas reverse happens in case of mushrooms. Mushrooms can be used as salad after light sautéing, used for making soups, snacks or vegetables. Most of the recipes available are or western type. We here describe some of the simple recipes that can be suitable for our conditions.

Mushroom as Salad

Clean button mushrooms thoroughly and cut these into flat thin pieces. These can be directly used as salad or sautéed a little. For the latter, put little oil on tawa, lightly fry mushrooms, sprinkle black pepper and lemon juice before serving it as salad.

Mushroom Tomato Soup

Ingredients

The following is required for preparing mushroom tomato soup

1. Fresh mushrooms(button or oyster) - 200 g (chopped)
2. Tomatoes - 4 (chopped)
3. Onion - 1 medium (chopped)
4. Garlic - 3 cloves
5. Corn flour - 3 table spoon
6. Cream - 2 table spoon
7. Butter - 50 g
8. Salt and pepper - to taste

Method

Boil tomatoes, onion and garlic in water for 10 minutes. Grind the above material and sieve it. Melt butter in a pan and sauté chopped mushrooms for 10 minutes until they become golden brown. Add sieved material to it, then add corn flour and boil the mixture for few minutes till it thickens. Serve hot with cream.

Mushroom Pakoda

Ingredients

The following is required for preparing mushroom pakoda

1. Fresh mushrooms - 500 g (Button or oyster or milky)
2. Gram flour (besan) - 150 g
3. Onion - 1 big (chopped)
4. Ginger - 2 tbl (chopped)

5. Garam masala - 10 g
6. Anar dana powder - 1 tbl.
7. Cooking oil - 100 g
8. Salt, green chillies (chopped) - to taste

Method

It is good to blanch the mushrooms before using these for making pakoda. Boil the mushrooms in hot water containing some salt for 5 minutes, remove excess water by sieving, allow these to dry and then cut and mix with other ingredients mentioned above. Mix well. Make a mixture of gram flour and add salt and chillies to it as well. Make small balls of the mixture, dip in besan and fry till golden. Serve hot with chutney or sauce.

We can make various other types of pakodas. These include use of small mushrooms as such, use of mushroom caps joined from gill side after filling cheese or palak inside and joined using tooth pricks (mushroom duplex). We can splice 100g oyster mushroom into small pieces manually or with knife, and then add pinch of salt, pepper, etc and one or two spoon of besan. Mix it with hand. There is no need to add water. We can fry these directly till golden brown to get crisp pakoda.

Mushroom with Rajmah Curry

Ingredients:

Mushroom - 500gm	Green chilli - 50 gm	Garam masala - 2tsp
Rajmah - 250gm	Garlic - 1 pod	Chilli powder - 1 tsp
Tomato - 250gm	Ginger paste - 1Tbsp	Oil - 5 Tbsp
Onion - 100gm	Turmeric powder - 2 Tsp	Salt - to taste

Method of preparation:

- ❖ Wash and soak rajmah overnight and cook till lender.
- ❖ Wash and cut the mushroom in the desired size.
- ❖ Heat oil, fry chopped onion, garlic, ginger paste, tomato and mushroom till the water dries up.
- ❖ Add cooked rajmah and simmer for few minutes.
- ❖ Add turmeric powder, garam masala powder, chilli powder and salt.
- ❖ Cook for few minutes and garnish with coriander leaves.

Mushroom with Fish and Capsicum Fry

Ingredients:

Mushroom - 500gm	Ginger paste - 1tsp	coriander leaf - 50 gm
Fish - 300gm	Turmeric powder - 2 tsp	Garlic - 1 pod
Tomato - 200gm	Garam masala - 2tsp	Salt - to taste
Onion - 100gm	Chilli powder - 1 tsp	
Capsicum -200 gm	Oil - 200ml	

Method of preparation:

- ❖ Wash and fry fish pieces till slightly brown.
- ❖ Wash and cut the mushroom in the desired size.
- ❖ Heat oil, fry chopped onion, garlic, ginger paste, tomato and mushroom till the water dries up.
- ❖ Add fried fish and simmer for few minutes.
- ❖ Add turmeric powder, garam masala powder, chilli powder and salt.
- ❖ Cook for few minutes and garnish with coriander leaves.

Chilli Mushroom

Ingredients:

Mushroom - 500gm	Garlic - 1 pod	Coriander leaf -100gm
Capsicum - 300gm	Ginger paste - 1tsp	Oil - 200ml
Tomato sauce - 200ml	Garam masala - 2 tsp	Salt - to taste
Onion - 200gm	Chilli powder - 1 tsp	

Method of preparation:

- ❖ Wash and cut the capsicum in to small pieces.
- ❖ Wash and cut the mushroom in the desired size.
- ❖ Heat oil, fry chopped onion, garlic, ginger paste, and mushroom till the water dries up.
- ❖ Add capsicum and fry for few minutes and stir properly.
- ❖ Add tomato sauce, garam masala powder, chilli powder and salt.
- ❖ Cook for few minutes and garnish with coriander leaves.

Mushroom with Potato Fry

Ingredients:

Mushroom - 500gm	Chilli powder - 10gm	Salt - to taste
Potato - 3 large potato	Garam masala - 10gm	Coriander leaf - 50gm
Onion - 2 medium	Oil - 100ml	
Garlic and Ginger paste - 3 tsp	Soya sauce - 100 ml	

Method of preparation:

- ❖ Wash and cut onion, potato and mushroom.
- ❖ Heat oil and stir fry onions, then garlic and ginger paste.
- ❖ Add mushroom and potato and fry till it is properly cook.
- ❖ Add chilli powder and garam masala.
- ❖ Add soya sauce and continue to fry for 2 minutes, add salt to taste.

❖ When done, garnish with coriander leave.

Arleen and Alan Bessette's Dragon's Mist Soup

1 cup thinly sliced fresh Shiitake mushrooms

1 tablespoon vegetable oil

1 (14 ½-ounce) can chicken broth

1½ cups water

4 cloves garlic, minced

2 scallions with tops, minced

½ cup drained, rinsed, 2- to 4-inch long bamboo shoots

2 tablespoons soy sauce

¼ teaspoon white pepper

5 ounces tofu (cut into ½-1 inch cubes)

1 teaspoon salt

1 teaspoon sesame oil

In a saucepan, sauté mushrooms in vegetable oil over medium-low heat for approximately 5 minutes. Add broth, water, and garlic; bring to a boil, reduce heat and simmer 10 minutes. Add all remaining ingredients except sesame oil. Return to boil, reduce heat and simmer 5 minutes. Just before serving, stir in sesame oil. Serves

Jack Czarnecki's Shiitake in Burgundy Butter Sauce

½ cup chopped onions

3 tablespoons melted butter

1 cup water

½ teaspoon ground chili powder

1 teaspoon lemon juice

¼ cup red wine

1 teaspoon sugar

½ teaspoon ground coriander

1 teaspoon crushed fresh garlic

¼ teaspoon ground black pepper

1 teaspoon salt

1 tablespoon soy sauce

1 pound fresh Shiitake caps

1½ tablespoons cornstarch mixed with ⅓ cup water

In a skillet, sauté the onions in the butter until transparent, then add water. Add the other ingredients, except the mushrooms and cornstarch, and stir for 1 minute. Add the mushrooms, and turn the heat to low. Cover the skillet with a tight-fitting lid, and let simmer for 30 minutes. Thicken the mixture with the cornstarch-and-water mixture, and serve alone or over rice. According to Jack, the sauce actually enhances the flavor of the Shiitake. Serves 4.

Jack Czarnecki's Chicken with Oyster Mushrooms

1½ cups heavy cream

1 teaspoon crushed fresh garlic

1 tablespoon finely chopped onion

½ pound Oyster mushrooms, cut into 2-inch strips

Salt and pepper to taste

1 tablespoon cream sherry

1 tablespoon prosciutto

12 ounces cooked chicken or turkey meat from breast, sliced into 2-inch strips

½ cup chicken stock

2 tablespoons cornstarch mixed with ⅓ cup water

Combine all the ingredients except the cornstarch mixture in a heavy skillet. You may also want to save the salting until the dish is slightly heated. Heat until simmering over a medium flame, then continue simmering over a low flame for 5 minutes. Thicken with the cornstarch-and-water mixture, and adjust for salt as necessary. Serves 2.

Larry Lonik's Morel Quiche

¼ cup bacon (or "bacon bits")

1 pound Morels

½ cup chopped onion

½ cup chopped green pepper

1½ cups shredded baby

Swiss cheese

1½ cups milk

¾ cup Bisquick

3 eggs

1 teaspoon salt

¼ teaspoon pepper

Preheat oven to 400°F (200°C). In a 10-inch, lightly greased pie pan, mix bacon, mushrooms, onion, green pepper, and cheese. In medium-sized bowl, add milk, Bisquick, eggs, salt, and pepper. Beat until smooth. Pour into pie pan. Bake 35–40 minutes or until inserted toothpick comes out clean. Serves 2 to 4.

Irene and Gary Lincoff's Pickled Maitake (Hen-of-the- Woods)

1 pound Maitake mushrooms

Marinade:

1 cup white wine (dry)

⅓ cup olive oil

Juice from 1 lemon

A few slices of onion

A few sprigs of parsley

A few cloves of garlic

Salt, peppercorns, or other spices

Clean 1 pound of mushrooms and separate the "leaves" of Maitake into bite-sized pieces. Throw them in boiling water for 2 minutes, drain, and let cool on paper towels.

Combine marinade ingredients and season with a sprinkle of salt, peppercorns, bouquet garni, or other preferred spices. Bring marinade to a boil and cook the mushrooms in the marinade until done to taste (crunchy to soft).

Remove mushrooms. Discard solids from marinade. Store mushrooms covered with marinade in glass jars, with a thin layer of olive oil on top to help preserve. Store refrigerated. 4 to 8 servings.

Hope and Orson Miller's Hot Mushroom Dip Especial

1 pound fresh mushrooms

6 tablespoons butter

1 tablespoon lemon juice

2 tablespoons minced onion

1 pint carton low-fat sour cream

2 vegetable or chicken bouillon cubes (or 2 teaspoons granules)

Salt and pepper to taste

2 tablespoons soft butter or margarine

2 tablespoons flour

Chop mushrooms quite fine and sauté in pan with butter and lemon juice. Let simmer 5–10 minutes. Add onions, sour cream, bouillon, salt, and pepper. Simmer 5–10 minutes more. Make a paste of the remaining butter and flour. Add to hot mixture and stir until thickened. Serve hot, in fondue pot or chafing dish, with chips, crackers, or fresh vegetables. (Note: If thickened with seasoned breadcrumbs, fresh dill may be added as filling for Mushroom Squares. Use crescent roll dough. Pat the dough into a small 9-inchsquare baking pan, spread filling and cover with more dough. Bake 20–30 minutes at 375°F. Cut into squares and serve hot.) Hope has prepared this dish to the kudos of mycologists throughout the world. Serves 2 to 4.

Scott and Alinde Moore's Cheese-Mushroom Quiche

1 ready-made pie crust

1½ cups grated Swiss cheese

1 tablespoon butter

1 medium onion, chopped

¼ pound mushrooms, chopped

Dash of salt, pepper, and thyme

¼ teaspoon salt

1 ½ cups milk

¼ teaspoon dry mustard

4 eggs

3 tablespoons flour

Paprika

Cover bottom of pie crust with grated Swiss cheese. Meanwhile, sauté butter, onions, mushrooms, salt, pepper, and thyme. Add the sautéed ingredients to the pie crust. Beat together the remaining five ingredients and pour over mushroom layer in piecrust. Sprinkle with paprika. Bake at 375°F (190°C) for 40–45 minutes, or until center is firm. Serves 2 to 4.

Maggie Roger's Oyster Mushrooms with Basmati Rice and Wild Nettles

Serve over basmati rice, fluffy white rice, or lightly toasted sourdough bread cubes. Serve with steamed wild nettles or freshly steamed asparagus on the side and a clear light wine. "Friends will chirp with satisfaction once they get over the wildness of it all." Serves 2 to 3.

6 dozen nettles or 1 loose, full grocery bag

Salt and pepper to taste

1½ cups chopped Oyster (or Shimeji) mushrooms

1 tablespoon butter or olive oil

1 thinly sliced scallion

¼ cup thinly sliced celery

¼ cup chopped carrot

¼ teaspoon basil

¼ teaspoon thyme

¼ cup very dry sherry

1 cup chopped cooked turkey breast

Bouillon granules

½ cup cold water

2 tablespoon flour, or 1 tablespoon cornstarch

1 cup half-and-half cream

From Maggie's notes: At the time you're finding spring Oyster mushrooms, the wild nettles are about to flower and ready for picking. Use leather gloves or native wisdom to keep from getting stung. Break off just the top three leaf levels of each stalk. Prepare them by rinsing in cold water, steaming over boiling water for 7 minutes and lightly salting and peppering them. (You can pick nettles in the spring, blanch for 6 minutes, and freeze several packages for future use.) If served as below, no melted butter is needed; they have their own flavor.

Sauté mushrooms in butter or olive oil with scallion, celery, and carrots. Let simmer in its juices for 10 minutes or until slightly reduced. Add basil, thyme, sherry, turkey bits, bouillon granules, and simmer for another 10 minutes. Taste, add salt and freshly ground pepper as needed.

Make a thickening of the water and flour (or cornstarch). Stir into mushroom mixture slowly and let simmer for another 5 minutes or more, or until thickened, stirring occasionally. Just before serving, add the half-and-half, simmer for 5 minutes or so, then pour into a warmed bowl.

Robert Rosellini's Broiled Rockfish in an Oyster or Shimeji Mushroom and Ginger Sauce

The following recipe uses a yellow-eye rockfish from the Pacific Northwest, which is firm and delicately textured with a low fat content. The absence of fat in this particular fish provides its "clean" and pure quality. The following recipe is quick, simple, and easy for home preparation.

6 ounces fillet of rockfish

Olive oil

1 ounce white wine

¼ pound fresh, thinly sliced

Oyster or Shimeji mushrooms

½ ounce preserved ginger

1 ounce sweet butter (unsalted)

Brush the fillet of fish with oil and broil (or bake at 450°F) until the translucent flesh turns opaque, 7–10 minutes. Remove fish from broiling pan, place the same pan on oven burner at high heat, and deglaze pan with white wine. When wine has been reduced by twothirds, add sliced mushrooms and preserved ginger. Sauté for about 1 minute, add butter, and swirl ingredients together until butter forms a smooth texture, about 16–20 seconds. Remove immediately and pour over fish fillet.

Note: Achieving the smooth texture with butter requires practice to avoid breaking the butter. Once this simple technique is accomplished, there are many variations of this recipe procedure.

Butter contains half the fat content of most oils, and thus these recipes should be consistent with low-fat dining strategies. Serves 2.

Paul and Dusty's Killer Shiitake Recipe

Once in a while, you come across a simple recipe that elicits enthusiasticexclamations of joy. Recently we came up with one. We get inundated with requests for it. So here it is.

1 tablespoon olive oil

2 tablespoons sesame oil

1 tablespoon tamari or soy sauce

2–3 tablespoons white wine

Pinch of black pepper

1–2 cloves of crushed garlic

1 pound whole, fresh Shiitake mushrooms

Mix the oils, tamari, wine, and spices in a small bowl. Stir vigorously as the ingredients tend to separate. Set aside.

Cut the mushroom stems from the caps. Place gills facing up. Do not slice mushrooms! (The stems can be dried and used for a soup base or discarded.) Baste the sauce onto the gills of the mushrooms, making sure the gills become saturated with the sauce.

In a 350°F oven, bake mushrooooms uncovered for 30–40 minutes. Or you can barbecue on an open grill. The smoky flavor makes it even better. Serve hot with seafood, rice, pasta, or whatever. Unbelievably good. Yum! Serves 2–4.

Fresh Shiitake Omelet

8 eggs

¼ cup water

2 tablespoons tamari

½ pound fresh sliced Shiitake

2 tablespoons canola oil

½ small onion, chopped

2 cloves garlic, chopped

½ cup cashews

1 cup grated cheddar cheese

Salt and pepper to taste

Mix the eggs and water in a large bowl and beat thoroughly. Add tamari to mushrooms and sauté in a frying pan with oil until water is cooked out. Add onions and garlic, cook for 1 minute, then add cashews. In a medium-sized skillet, spray a thin coat of oil or butter, add egg mixture and cook for 1 minute. Add layer of cheese. Pour mushroom mixture over the cheese. Fold over and cook for 2–3 minutes. Add salt and pepper to taste. Serves 4.

Larry Stickney's Morel Crème Superieur

Larry's recipe is quick and easy, ideal for tasting Morels at primitive campsites during remote Morel forays in the mountains. "Only the medically forbidden should entertain thoughts of dietary restraint during the brief Morel season when long, hard hours in the forest set up a healthy appetite in the wake of massive calorie burning and likely liquid deprivation.... Delectation is soon at hand." He makes no note for those athletically inclined Morel cultivators who must walk from their house to their backyard to harvest Morels. Presumably, the same advice holds.

1–2 pints heavy cream

2–3 tablespoons butter

½ pound (or more) fresh Morels (cut into ¼-1 inch cartwheel sections)

Warm cream and butter over a camp stove or wood fire in a small pot. Do not let cream burn onto the pan's bottom. When the cream simmers, place the sliced Morels into pot, and cook until cream returns to a simmer for at least 5 minutes. Morels should not become limp before tasting, using toothpicks or forks to retrieve them. After the last Morel is removed, Larry notes that lots must be drawn to determine the lucky soul who has the privilege of downing the heavenly, rich, spore-darkened cream soup. Serves 2 to 4.

Andrew Weil's Shiitake Teriyaki

1 cup dried Shiitake

Hot water

¼ cup sake

¼ cup soy sauce

2 tablespoons light brown sugar

2 chopped scallions

A few drops (roasted) sesame oil

Reconstitute 1 cup dried Shiitake by covering with hot water and let stand till caps are completely soft. (Or cover with cold water, microwave on high for 2 minutes, and let stand.) Cut off and discard stems. Squeeze excess liquid from caps and slice into ¼-inch pieces.

Place pieces in saucepan with sake, soy sauce, and brown sugar. Bring to boil and simmer, uncovered, till liquid is almost evaporated, tossing mushrooms occasionally. Remove from heat, cool, and chill. Sprinkle with finely chopped scallions and a few drops of dark (roasted) sesame oil. Serve as appetizer, side dish, or over rice.

Shiitake or Maitake Clear Soup

½ ounce dried mushrooms (or ¼ pound fresh mushrooms)

2 cups water

1 teaspoon soy sauce

2 tablespoons miso

1 pound tofu, coarsely chopped

¼–½ cup chopped onions

Soak cut or broken dried mushrooms in 2 cups cold water for 15 minutes. Drain off broth and save. Cover mushrooms with more cold water and soak for another 20 minutes. Add saved mushroom broth back into preparation and boil it for a few minutes. Season soup with soy sauce and miso. Add tofu and onions. (If needed, add 1–2 cups more water.) Bring back to a boil for 1–2 minutes and turn down heat. Allow to simmer for 5 minutes. Serves 4.

Shiitake Hazelnut Vegetarian Pâté

4 ounces Shiitake mushrooms

3 tablespoons butter

1 clove garlic, minced

⅛ teaspoon thyme

¼ teaspoon salt

⅛ teaspoon pepper

1 teaspoon fresh parsley leaves

¼ cup toasted hazelnuts

3 ounces Neufchâtel cheese

2 teaspoons dry sherry

Trim and discard woody ends from mushrooms. In a food processor, finely chop mushroom caps and stems. In medium skillet, melt butter. Add mushrooms and garlic and sauté for at least 5 minutes.

Stir in thyme, salt, and pepper. In food processor, chop parsley. Add hazelnuts and process. Add Neufchâtel cheese and process until smooth. Add sherry and mushroom mixture. Process until well mixed. Spread or mold in serving dish. Cover. Chill at least 1 hour.

Serve with crackers. Yield: 1 cup. (Other mushrooms can be substituted for or combined with Shiitake.)

Maitake "Zen" Tempura

This recipe can also be used as a base to make tempura shrimp, whitefish, and/or assorted vegetables (zucchini, potatoes, onions, etc.).

1 ounce dried or ⅓ pound fresh Maitake

2 cups flour (⅓ pound)

2 eggs

2 cups plus ½ cup cold water

Vegetable or canola (rapeseed) oil for frying

Tempura sauce

If using dried Maitake, soak cut or broken mushrooms in 2 cups cold water for 15 minutes.

Discard water. Cover mushrooms with more cold water and soak for another 20 minutes. Drain and discard water. In a separate bowl, mix flour with eggs and cold water. Dip and roll the mushrooms into the flour/egg mixture. Deep-fry the Maitake mushrooms in hot oil (356°F / 180°C) for 1 minute. Remove any damp excess oil with paper towel. Serve with tempura sauce. Serves 2.

Stuffed Portobello Mushrooms

8 large portobello mushrooms

1 tablespoon olive oil

2 red bell peppers, seeded and finely chopped

2 green bell peppers, seeded and finely chopped

1 large onion, finely chopped

3 scallions (spring onions), green and white parts, thinly sliced

5 to 8 cloves garlic, finely chopped, to taste

½ teaspoon oregano

½ teaspoon dried basil

½ teaspoon thyme

Salt and freshly ground pepper to taste

6 ounces goat's cheese (optional)

Additional sliced scallions, for garnish

Remove the stems from the portobellos, chop and reserve. Place the whole mushroom caps smooth side down on a lightly greased baking sheet and bake in a preheated 425°F (220°C) oven for 15 minutes. Meanwhile, heat the oil in a skillet over moderate heat and sauté the mushroom stems, bell peppers, onion, scallions, and garlic until tender, 8 to 10 minutes. Add the herbs and cook an additional 2 minutes. Spoon the vegetable mixture into the mushroom caps and top with the cheese. Bake an additional 10 minutes, or until the mushrooms are tender and the cheese has melted. Sprinkle with sliced scallions and serve immediately. Serves 4 (2 mushrooms per person).

Mushroom as Vegetable

Preparing mushroom vegetable is similar to making any other vegetable. One major difference is that on over cooking other vegetables may become soft, but mushrooms tend to become harder. Prepare masala using onion, garlic, tomato, turmeric powder, garam masala, red chillies, dhania powder, salt, etc as we do for any other vegetable. Add green peas or Shimla mirch and after slight cooking add mushrooms. Mushrooms can be made without addition of any vegetables as well. Add water as per need depending upon the consistency of the gravy you prefer.

Quality Assurance, Shelf Life and Packaging

For successful marketing it is important that our product should good consistent quality with proper details of its nutrition, date of packaging, etc. It is also important that our packaging should be attractive and suitable for the product being marketed. Here are some of the aspects you should take care of to ensure value for the money.

Quality Assurance

Quality of processed product is very important for marketing it. The concept of Total Quality Management (TQM) should be applied if we are going to do commercial production. It means right first time, every time. That is, our every product should be of the same quality in every batch right from the first packet to the last packet. We cannot provide different quality and our product should taste same every time. For example you must have tasted many soft drinks or chocolates. They taste same irrespective of when or from where you purchase.

This is so because the companies follow the concept of TQM.

Shelf Life

Processed products have long shelf life. The shelf life of all products however will not be same. You must have seen that whenever we purchase any processed product, there is clear mention of date of manufacture and date by which it can be used. There can be other instructions like store in cool and dry place, or use within how many days of opening.

Packing, Packaging and Market Opportunities

Proper packing is important to ensure that all packets have same shelf life. If we use poor quality bags or cans, then it can lead to loss of quality in some of the packets. Proper labelling with all the nutritional, information and other mandatory details is must. A good label is also important to attract the attention of the customer. Hence, it is important to use the same label to establish your brand in the market. The label may also have instruction regarding how to use the product. Many a times brief recipes can be mentioned to promote the product.

Conclusion

In the recent years, there has been an increasing interest in "value added" agricultural products in the market in the modern era. With so many fortified and value added products available in the market, inclusion of products made from mushrooms needs to be emphasized. Also, the retention of fresh mushrooms at various levels such as with growers, whole seller, retailers and consumers results in deterioration in the quality of the mushrooms and great economic loss. To overcome all these problems, adaption of appropriate post harvest technologies and processing of surplus mushrooms in the form of novel value added products isrequired. Although, mushrooms are considered rich source of protein, vitamins, minerals (Ca, P, K, Mg, Zn, Se etc) and nutraceuticals, the presence of strong smell due to the various volatile compounds is relished by some and not liked that much by others. The processing of these mushrooms into different products could be the novel initiative towards their consumption among all people in various age groups. Also, they can be easily stored and can be used as nutrient reservoirs to meet the daily requirements of nutrients.

References

1. Shrivastava, M. (1998). Studies on mushroom dehydration (*Pleurotus florida*). *PhD. Thesis submitted to IIT, KGP, W.B., India*

2. Chang, S. T. and J. A. Buswell. 1996. Mushroom nutriceuticals. *World J. Microbiology Biotech,*12:473-476.

3. Poongkodi, G. K. and D, Sakthisekaran. 1995. Nutrient content of the mushrooms. *Madras Agricultural Journal,* **82**(9):555-556.

4. Cheung, P. C. K. 1998. Functional properties of edible mushrooms. *Journal of Nutrition,*128:1512-1516.

5. Sharma, R. K. and Dhar, B. L. 2010. Mushroom cultivation: A highly remunerative crop for Indian farmers. *Indian Farming (New Direction),* January,2010.

6. Mandeel Q, Al-Laith A, Mohamed S. Cultivation of oyster mushrooms (Pleurotus spp.) on various lignocellulosic wastes. World Journal of Microbiology and Biotechnology. 2005;21(4):601-7.

7. Garcha H, Khanna P, Soni G. Nutritional importance of mushrooms. In: Chang S, Buswell J, Chiu S, editors. Mushroom biology and mushroom products Chinese University. Hong Kong: Chinese University Press; 1993.

8. Tibuhwa DD. Wild mushroom-an underutilized healthy food resource and income generator: experience from Tanzania rural areas. Journal of ethnobiology and ethnomedicine. 2013;9(1):49.

9. World Bank. Ethiopia: Country economic memorandum Geneva World Bank, 2004.

10. Dawit A. Mushroom Cultivation: A practical approach. Addis Ababa: Birhannaselam printing enterprise; 1998.

11. Oei P, van Nieuwenhuijzen B. Small-scale mushroom cultivation: oyster, shiitake and wood ear mushrooms. Wageningen, The Netherlands: Digigrafi; 2005.

Index

T

V

W